# MathWise Integers

## With Answer Key

## Peter Wise

MATH TEACHER,
MONUMENT,
COLORADO

CONTRIBUTORS

David Wise

Katherine Wise

Cover Design by Kris Budi

*Dedicated to the students in my after-school math clubs, whose commitment and interest in math motivated me to write this workbook.*

*...with special thanks to:*

- *Aileen Finnegan, principal, for her support and encouragement of the math clubs at our school*
- *Barbara Langlois, for her cheerful and efficient administrative help*
- *Shi Hayes, school office manager, for her logistical help with the math clubs*

**MathWise Integers with Answer Key**

MathWise Curriculum Press

First printing 2014

# MathWise Integers

## TABLE OF CONTENTS

## FOR WHICH GRADE LEVEL(S) ARE THESE BOOKS INTENDED?

This series is based on skill sets, not grade or age. These workbooks are intentionally created to be suitable for a wide range of grades. They are focused on math topics, irrespective of grade level or age. If students in any grade need extra support in a given topic area, these books are designed to enrich whatever curriculum they may be using. If a student in middle or high school student is rusty on a skill set normally taught in earlier grades, this series will help. There should be no stigma attached to reviewing important content in math, language arts, or even a foreign language. On the other hand, if a student in 3rd grade has a parent or teacher who is willing to walk the student through the explanations and exercises in this book, he or she will also profit from the study. Front-loading key math concepts will make future math classes that much easier.

These books were part of the instruction in several different grade levels, and even in multi-grade math clubs. No one gets distracted by the grade level of the material. The concepts are the target.

## MY EMPHASES IN TEACHING MATH

Too many students learn math as if they were learning a dead language. To them math consists of memorizing a bunch of rules and formulas. This is the wrong approach to learning math. To be good at math, it is important to know **how and why math works the way it does.** Students need to be trained to think mathematically from preschool through college, in every grade level. A formulaic understanding of math is both harder to learn and easier to forget.

Tips and tricks help as memory aids and have a legitimate role in acquiring and retaining information. However it is even more important that students understand the reasoning behind rules and formulas. **The MathWise series incorporates both tips/tricks as well as reasoning behind math formulas and procedures.**

## THE FORMAT OF THIS BOOK

The explanations, graphics, and format of this series is designed to be kid-friendly, upbeat, and as appealing as possible. I have incorporated tips, tricks, and other pedagogical secrets into this book. Students tell me that they like the format and self-contained explanations. Every year students make breakthroughs in their understanding of math through these pages.

## HOW TO USE THIS BOOK

While students can use this series profitably when working alone, my experience has indicated that they will make greater progress if guided by a parent, tutor, or helper. This is particularly true for younger students. This person need not be a math teacher at all—just a reader.

If a student or parent is unclear about a solution or procedure, he or she is encouraged to check with the answer key at the back of the book.

### Web Site
For questions, comments, or suggestions, please visit www.mathwisebooks.com.

## TEACHING INTEGERS

Most students find integers to be a bit confusing and counter-intuitive at the beginning stages. This is especially true with integer addition and subtraction (most students find integer multiplication and division to be much easier).

To help students with this topic, this book presents three different approaches to adding and subtracting integers:

• Number lines (both horizontal and vertical)
• Counters
• Rules: Same Sign Rule and the Opposite Sign Rule

At the beginning of their studies, students need to know that subtraction problems can be rewritten as adding negative numbers; i.e., 10 - 2 can be rewritten as 10 + (-2). The 2 that is being subtracted counts as a negative number.

Inequalities are presented because they help to visualize the relationships between negative and positive integers.

This series of books is designed to be unique and to catch kids' attention in special ways:

**Tips and Tricks**
Over the years, I have assembled a wide assortment of memory aids—my tips and tricks. Students have found these to be helpful and memorable, but they have also found that these pointers add interest and excitement to their math studies.

**Speech Bubbles with Teacher Insights**
Speech bubbles are used to provide guidance, point out insights, or give helpful hints as students are solving math problems. Students learn best by doing, and the instruction given in the speech bubbles is designed to (1) sharpen students' powers of observation, (2) increase number sense, and (3) instruct in digestible chunks.

**Simplicity of Instruction**
Concepts are explained clearly and simply. Every page (excluding review pages or concept quizzes) has a specific focus. Most pages have generous amounts of white space to keep students focused. Movement is from the simple to the increasingly complex.

**Step-By-Step Procedures**
Students learn best when given explicit, step-by-step instruction. When several steps are involved, they are numbered. This makes learning much more logical and memorable.

**Depth and Complexity**
Throughout the book there are challenge problems to stretch students' thinking, and they are generally labeled as such. At your discretion, you can guide students through the more challenging problems (recommended) or, alternatively, you can skip these harder problems.

**Informal Terms**
This book often employs informal language like "top number" or "bottom number" to keep things simple and focused. Standard mathematical terminology, such as numerator and denominator, is used after the concepts are presented.

**Logical-Sequential Instruction**
Math problems are presented in a logical sequence so that previous problems contribute to students' abilities to solve future problems. The order in which you present math problems is critical to promoting number sense.

**Concept Quizzes**
Concept quizzes focus on the methodology and on students' ability to articulate the approach to various kinds of problems.

# What Are Integers Anyway?

Integers are whole numbers and their opposites.

> NO BETWEEN NUMBERS! (NO FRACTIONS OR NUMBERS TO THE RIGHT OF THE DECIMAL POINT)

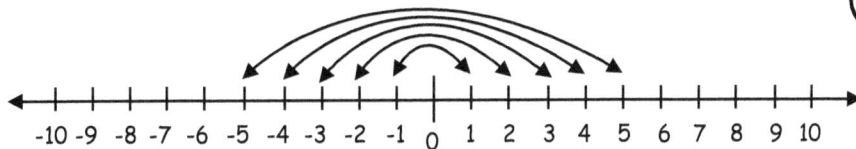

-10 -9 -8 -7 -6 -5 -4 -3 -2 -1 0 1 2 3 4 5 6 7 8 9 10

> POSITIVE NUMBERS ARE USUALLY NOT WRITTEN WITH A POSITIVE SIGN (+2, ETC.)!

## Circle the integers (positive or negative numbers—no between fractions or decimals)

1. $7$ $\frac{1}{2}$ $.3$ $-9$

2. $-.4$ $\frac{3}{4}$ $-2$ $+5$

3. $-3\frac{1}{2}$ $12$ $-6$ $+\frac{6}{5}$

## Give the opposite of each integer

4. $7$ ☐

> SAME NUMBER, JUST PUT ON A NEGATIVE SIGN...THE OPPOSITE OF POSITIVE IS MINUS!

5. $-3$ ☐

> THE OPPOSITE OF A NEGATIVE IS A POSITIVE, SO JUST TAKE OFF THE SIGN!

6. $-5$ ☐

7. $-x$ ☐

8. $y$ ☐

8. $-a$ ☐

## Put a dot on each number and its opposite on the number line; connect them with arrows

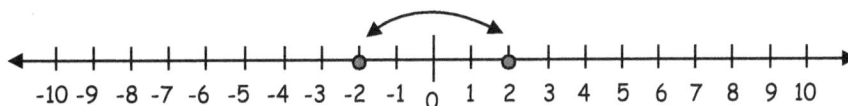

-10 -9 -8 -7 -6 -5 -4 -3 -2 -1 0 1 2 3 4 5 6 7 8 9 10

9. $2$

10. $-4$

11. $6$

12. $-10$

This one is done for you

1

# Subtraction as Adding Negatives

Any time you subtract, this is the same as adding a negative number

$$10 - 3 \longrightarrow 10 + (-3)$$

'MINUS' IS THE SAME AS 'PLUS A NEGATIVE'!

REWRITE −3 AS + (−3)

EVERY TIME YOU SUBTRACT, IT'S THE SAME AS ADDING A NEGATIVE NUMBER!

---

**Rewrite the following subtraction problems as adding negative integers**

1. $5 - 2 \longrightarrow$ [ + (−  ) ]   (Right now just focus on rewriting, rather than solving the problem)

2. $6 - 5 \longrightarrow$ [ + (−  ) ]

3. $7 - 4 \longrightarrow$ [        ]

4. $12 - 6 \longrightarrow$ [        ]

5. $10 - 3 \longrightarrow$ [        ]

6. $14 - 7 \longrightarrow$ [        ]

7. $20 - 5 \longrightarrow$ [        ]

8. $15 - 10 \longrightarrow$ [        ]

9. $-16 - 6 \longrightarrow$ [        ]

10. $-8 - 2 \longrightarrow$ [        ]

11. $13 - 7 \longrightarrow$ [        ]

12. $19 - 11 \longrightarrow$ [        ]

13. $-30 - 5 \longrightarrow$ [        ]

14. $a - b \longrightarrow$ [        ]

15. $x - y \longrightarrow$ [        ]

---

BACKWARDS:
Now write as a subtraction problem

16. $a + (-b) \longrightarrow$ [        ]

17. $\frac{2}{3} + \left(-\frac{1}{2}\right)$ [        ]

2

# Same Sign Rule

SAME SIGN RULE
"SSR"

- ADD THE NUMBERS
- ADD THE SIGN!

IF ALL THE NUMBERS ARE NEGATIVE, JUST ADD UP ALL OF THE NUMBERS AND PUT ON A NEGATIVE SIGN!

IF ALL THE NUMBERS HAVE THE SAME SIGN, JUST GIVE YOUR ANSWER THE SAME SIGN AND ADD UP ALL THE NUMBERS!

With same signs you always ADD the numbers

2 NEGATIVE ONES + 3 NEGATIVE ONES = HOW MANY NEGATIVE ONES?

## Add or subtract the following integers

1. $-2 + (-3) = -\boxed{\phantom{0}}$

JUST ADD UP THE NUMBERS AS THOUGH THEY ARE POSITIVE, AND MAKE YOUR ANSWER NEGATIVE AT THE END!

SINCE BOTH NUMERS ARE NEGATIVE, YOUR ANSWER WILL BE NEGATIVE!

7. $-5 - 1 = \boxed{\phantom{0}}$

BOTH INTEGERS ARE NEGATIVE SO THE ANSWER WILL BE NEGATIVE.

4 + 6 GOES HERE (YOU KNOW THAT THE ANSWER IS NEGATIVE BECAUSE EVERYTHING YOU ARE ADDING IS NEGATIVE!

2. $-4 + (-6) = -\boxed{\phantom{0}}$

8. $-10 + (-2) = \boxed{\phantom{0}}$

3. $-5 + (-2) = -\boxed{\phantom{0}}$

IF YOU TAKE AWAY 5 (−5) AND YOU TAKE AWAY 2 (−2) HOW MANY DID YOU REALLY TAKE AWAY?

4. $6 + 3 = +\boxed{\phantom{0}}$

9. $-1 + (-1) + (-1) = \boxed{\phantom{0}}$

THIS TIME BOTH NUMBERS ARE POSITIVE, SO THE ANSWER WILL BE POSITIVE!

10. $-2 + (-2) + (-2) = \boxed{\phantom{0}}$

5. $-7 + (-7) = \boxed{\phantom{0}}$

11. $-7 - 2 - 3 = \boxed{\phantom{0}}$

6. $-7 - 8 = \boxed{\phantom{0}}$

THIS IS REALLY THE SAME AS SUBTRACTING 15 IN TWO STAGES

12. $-4 - 8 + (-2) = \boxed{\phantom{0}}$

3

# You Always Add Same Sign Integers

If integers have the same sign (either positive or negative) you ALWAYS ADD them. After you have added them, just make sure you PUT ON A NEGATIVE sign if the numbers you added are NEGATIVE.

(You don't normally put a positive sign on positive numbers. We assume that if a number has no sign, it's positive.)

### Add negatives together like they're positive; then put on a negative sign

A. -3 - 3 =

Both numbers are negative

Put on the negative sign: □    □  Add them like they're positive

### Smaller negatives when added equal a larger negative

---

**ADD these SAME SIGN integers; remember to put on the right sign!**

1. -4 - 4 = □    *IF YOU TAKE AWAY 4 TWICE YOU GET THIS!*

2. 4 + 4 = □    *IF YOU ADD 4 TWICE YOU GET THIS!*

3. -4 + (-10) = □

4. (+7) + (+8) = □

5. (-5) + (-1) = □    *TAKE AWAY ONE MORE THAN 5!*

6. (-5) + (-5) = □    *SAME AS 2 TIMES (-5)!*

13. 6 + 7 = □

14. (-8) - 10 = □

15. (-2) - 3 = □

16. -2 + (-14) = □

17. 17 + 3 = □

18. -9 + (-3) = □

# Same Sign Integers

1. -5 - 5 = ☐

2. -3 + (-6) = ☐

3. (-2) + (-5) = ☐

4. 3 + 8 = ☐

5. (+4) + (+6) = ☐

6. (-2) + (-2) = ☐

7. 17 + 17 = ☐

8. (-17) + (-17) = ☐

9. 14 + (+3) = ☐

10. -2 - 13 = ☐

11. -4 + (-4) = ☐

12. 12 + (+13) = ☐

13. -10 + (-4) = ☐

14. (-5) - 7 = ☐

15. 14 + 8 = ☐

16. -12 + (-2) = ☐

17. 16 + 5 = ☐

18. (-7) - 9 = ☐

19. 10 + 14 = ☐

20. -4 - 16 = ☐

21. 135 + 135 = ☐

22. -100 + (-47) = ☐

23. -65 + (-20) = ☐

24. -135 - 20 = ☐

# Combining Negative Numbers

Example

A group of SMALLER negative numbers can be combined to make one BIGGER negative number!

This is really just subtracting the same amount in two stages!

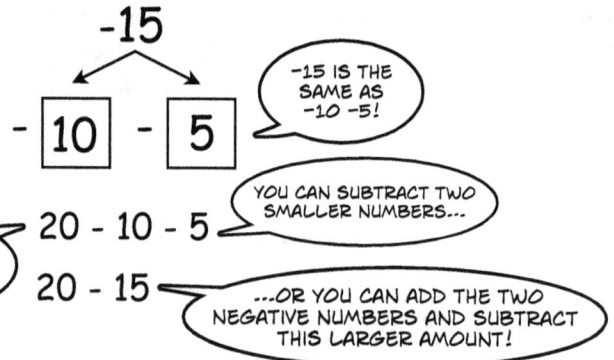

$$-15$$

$$- \boxed{10} - \boxed{5}$$

*−15 IS THE SAME AS −10 −5!*

*YOU CAN SUBTRACT TWO SMALLER NUMBERS...*

$$20 - 10 - 5$$

*THESE ARE THE SAME!*

$$20 - 15$$

*...OR YOU CAN ADD THE TWO NEGATIVE NUMBERS AND SUBTRACT THIS LARGER AMOUNT!*

---

## Break up the negative number into smaller negative numbers

**1.** $-5$

$8 - \boxed{3} - \boxed{\phantom{0}} = \boxed{\phantom{0}}$

**4.** $-20$

$25 - \boxed{\phantom{0}} - \boxed{\phantom{0}} = \boxed{\phantom{0}}$

**7.** $-9$

$12 - \boxed{\phantom{0}} - \boxed{\phantom{0}} = \boxed{\phantom{0}}$

**2.** $-7$

$10 - \boxed{2} - \boxed{\phantom{0}} = \boxed{\phantom{0}}$

**5.** $-12$

$18 - \boxed{\phantom{0}} - \boxed{\phantom{0}} = \boxed{\phantom{0}}$

**8.** $-14$

$16 - \boxed{\phantom{0}} - \boxed{\phantom{0}} = \boxed{\phantom{0}}$

**3.** $-10$

$20 - \boxed{\phantom{0}} - \boxed{\phantom{0}} = \boxed{\phantom{0}}$

*PICK ANY TWO NUMBERS THAT ADD UP TO 10!*

**6.** $-6$

$10 - \boxed{\phantom{0}} - \boxed{\phantom{0}} = \boxed{\phantom{0}}$

**9.** $-10$

$14 - \boxed{\phantom{0}} - \boxed{\phantom{0}} - \boxed{\phantom{0}}$

$= \boxed{\phantom{0}}$

*PICK ANY THREE NUMBERS THAT ADD UP TO 10!*

# Making One Large Negative Number

When you have a group of negative numbers, add them up and make one BIG negative!

*ADD THEM LIKE THEY ARE POSITIVE, BUT PUT ON A NEGATIVE SIGN!*

10 [-2  -2  -2]

*IF YOU SUBTRACT 2 THREE TIMES, IT IS THE SAME AS SUBTRACTING 6!*

10  −  6

Remember that -a and + (-a) are the same!

---

**Combine the smaller negatives into one big negative number; then solve**

**1.**   -5 - 4

*ADD UP THE NEGATIVES LIKE THEY ARE POSITIVE...JUST PUT ON THE NEGATIVE SIGN AFTERWARDS!*

= -  ☐

*TWO SMALLER NEGATIVES MAKE ONE LARGER NEGATIVE!*

10 - 5 - 4

= 10 - ☐ = ☐

**2.**   -3 - 3

= -  ☐

10 - 3 - 3

= 10 - ☐ = ☐

**3.**   -5 - 5 - 2

= -  ☐

20 - 5 - 5 - 2

= 20 - ☐ = ☐

**4.**   10 - 2 - 2

10 - ☐ = ☐

**5.**   16 - 1 + (-1)

16 - ☐ = ☐

**6.**   7 - 1 - 2

7 - ☐ = ☐

**7.**   12 - 2 + (-3)

12 - ☐ = ☐

**8.**   13 + (-3) + (-6)

13 - ☐ = ☐

7

# Same Sign Integers

| Solve these same sign integer problems |
| --- |

**1.**   8 + 8 =  ▢

**2.**   -8 + (-8) =  ▢

_____8_____ neg steps + _____ neg steps = _____ neg steps

**3.**   (-3) + (-1)  =  ▢

take away ___3___ and take away _____

= take away _____

**4.**   2 + 3 =  ▢

add _____ and add _____

= _____

**5.**   (+7) + (+3) =  ▢

**6.**   (-4) + (-4) =  ▢

**7.**   -2 + (-2) =  ▢

**8.**   16 + 16  =  ▢

**9.**   (-16) + (-16) =  ▢

**10.**   13 + (+15) =  ▢

**13.**   -5 + (-3)  =  ▢

___5___ steps LEFT + _____ steps LEFT = _____ steps LEFT
on a number line    on a number line    on a number line

**14.**   (-6) - 1  =  ▢

**15.**   8 + 9  =  ▢

**16.**   -1 + (-5)  =  ▢

**17.**   14 + 7  =  ▢

**18.**   (-12) - 8  =  ▢

**19.**   13 + 7  =  ▢

**20.**   -8 - 5  =  ▢

**21.**   -50 - 50  =  ▢

# Calculating with Groups of Negatives

**Example**

**A.** $9 - 2 - 3 = \boxed{4}$

minus 5

pos. → $\boxed{9}$

neg's → $\boxed{-5}$

YOU CAN SUBTRACT IN TWO STEPS (-2 AND THEN -3), BUT HERE COMBINE BOTH NEGATIVES TO MAKE ONE LARGER NEGATIVE!

REMEMBER! IF IT DOESN'T HAVE A SIGN, IT'S POSITIVE!

## Add/subtract the following integers

**1.** $10 - 3 - 1 = \boxed{\phantom{0}}$

(10)  pos. → $\boxed{\phantom{0}}$

(-3 - 1)  neg's → $\boxed{\phantom{0}}$

ADD THE NEGATIVES TO MAKE ONE LARGER NEGATIVE!

Add negatives like the are positive, then put on a negative sign

**2.** $-3 + 20 - 2 = \boxed{\phantom{0}}$

pos. → $\boxed{\phantom{0}}$

neg's → $\boxed{\phantom{0}}$  (-3 - 2)

**3.** $-2 - 2 - 2 + 18 = \boxed{\phantom{0}}$

pos. → $\boxed{\phantom{0}}$

neg's → $\boxed{\phantom{0}}$

**4.** $-4 + 16 + (-7) = \boxed{\phantom{0}}$

pos. → $\boxed{\phantom{0}}$

neg's → $\boxed{\phantom{0}}$

**5.** $-5 + 14 - 3 = \boxed{\phantom{0}}$

pos. → $\boxed{\phantom{0}}$

neg's → $\boxed{\phantom{0}}$

**6.** $-6 - 3 + 12 = \boxed{\phantom{0}}$

pos. → $\boxed{\phantom{0}}$

neg's → $\boxed{\phantom{0}}$

**7.** $20 - 2 - 5 = \boxed{\phantom{0}}$

pos. → $\boxed{\phantom{0}}$

neg's → $\boxed{\phantom{0}}$

**8.** $-8 + 15 - 2 = \boxed{\phantom{0}}$

pos. → $\boxed{\phantom{0}}$

neg's → $\boxed{\phantom{0}}$

**9.** $-7 + 18 - 7 = \boxed{\phantom{0}}$

pos. → $\boxed{\phantom{0}}$

neg's → $\boxed{\phantom{0}}$

**10.** $15 - 3 - 4 = \boxed{\phantom{0}}$

pos. → $\boxed{\phantom{0}}$

neg's → $\boxed{\phantom{0}}$

# Calculating with Groups of Negatives, pt. 2

## Add/subtract the following integers

**1.**  30 - 5 - 10 = ☐

pos. → ☐

neg's → ☐

*ADD THE NEGATIVES TO MAKE ONE LARGER NEGATIVE!*

**2.**  -2 + 28 - 4 = ☐

pos. → ☐

neg's → ☐

**3.**  -7 + 19 + (-5) = ☐

pos. → ☐

neg's → ☐

**4.**  -5 - 9 + 17 = ☐

pos. → ☐

neg's → ☐

**5.**  -6 + 14 + (-7) = ☐

pos. → ☐

neg's → ☐

**6.**  24 + (-8) + (-4) = ☐

pos. → ☐

neg's → ☐

**7.**  15 + (-5) + (-6) = ☐

pos. → ☐

neg's → ☐

**8.**  (-6) + 11 + (-7) = ☐

pos. → ☐

neg's → ☐

**9.**  -2 + (-4) + (-3) + 14 = ☐

pos. → ☐

neg's → ☐

**10.**  -3 + 2 - 8 + 4 = ☐

pos. → ☐

neg's → ☐

**11.**  -2 + (4) - 6 + (10) = ☐

pos. → ☐

neg's → ☐

**12.**  (3) - 5 + (8) - 9 = ☐

pos. → ☐

neg's → ☐

# Concept Quiz

1. Integers are _____ numbers and their _____ .

2. Whenever you SUBTRACT, it's the same as _____ing a _____ number.

3. Rewrite 10 - 6 as an addition problem: _____

4. If integers have SAME SIGNS (all positive numbers or all negative numbers) you always ADD / SUBTRACT the numbers before putting on the same sign.

   (circle one)

5. When adding a group of smaller negative numbers you always get a SMALLER / LARGER negative number.

   (circle one)

6. Rewrite -4 - 4 as an addition problem: _____  Solve it: _____

7. Rewrite 10 - 2 - 6 as a problem with only two numbers: _____

8. If you took away 3 apples and took away another 2 apples how many apples did you really take away?

   _____

# Opposite Sign Rule

© Peter Wise, 2014

## Example

THE NUMBER WITH THE NEGATIVE SIGN IS GREATER THAN THE NUMBER THAT IS POSITIVE!

**A.** $-12 + 10$

Make both numbers positive: | 12 | 10 |

Find the difference (subtract): | 2 |

Find the sign of the higher number; give your answer this sign: | - |

Answer: | -2 |

## Follow the steps and fill in the boxes to get your answers

ADDITION RULE FOR OPPOSITE SIGN INTEGERS:

*PRETEND THEY'RE POSITIVE...    SUBTRACT...    ADD THE WINNING SIGN!*

**1.** $6 + (-7)$

Make both numbers positive: | | |

Find the difference (subtract): | |

Find the sign of the higher number; give your answer this sign: | |

Answer: | |

AFTER YOU MAKE BOTH NUMBERS POSITIVE, SUBTRACT THE SMALLER NUMBER FROM THE LARGER NUMBER!

**2.** $-8 + 5$

Make both numbers positive: | | |

Find the difference (subtract): | |

Find the sign of the higher number; give your answer this sign: | |

Answer: | |

**3.** $-10 + 4$

Make both numbers positive: | | |

Find the difference (subtract): | |

Find the sign of the higher number; give your answer this sign: | |

Answer: | |

**4.** $7 - 12$

Make both numbers positive: | | |

Find the difference (subtract): | |

Find the sign of the higher number; give your answer this sign: | |

Answer: | |

**5.** $-6 + 17$

Make both numbers positive: | | |

Find the difference (subtract): | |

Find the sign of the higher number; give your answer this sign: | |

Answer: | |

# Opposite Sign Rule

|  | | Make both numbers positive | Find the difference (subtract) | Find the sign of the higher number; give your answer this sign | Answer: |
|---|---|---|---|---|---|
| 1. | -7 + 3 | ☐ ☐ | ☐ | ☐ | ☐ |
| 2. | 4 + (-9) | ☐ ☐ | ☐ | ☐ | ☐ |
| 3. | -14 + 10 | ☐ ☐ | ☐ | ☐ | ☐ |
| 4. | (-3) + 9 | ☐ ☐ | ☐ | ☐ | ☐ |
| 5. | -18 + 20 | ☐ ☐ | ☐ | ☐ | ☐ |
| 6. | 6 + (-9) | ☐ ☐ | ☐ | ☐ | ☐ |

ADDITION RULE FOR OPPOSITE SIGN INTEGERS:

PRETEND THEY'RE POSITIVE... SUBTRACT... ADD THE WINNING SIGN!

# Opposite Sign Rule

| | | Make both numbers positive | | Find the difference (subtract) | Find the sign of the higher number; give your answer this sign | Answer: |
|---|---|---|---|---|---|---|
| 1. | 6 + (-8) | | | | | |
| 2. | -6 + 8 | | | | | |
| 3. | -5 + 2 | | | | | |
| 4. | (-9) + 10 | | | | | |
| 5. | 12 + (-20) | | | | | |
| 6. | -30 + 24 | | | | | |
| 7. | 16 - 19 | | | | | |

14

# Adding Integers on a Number Line

**Example**

**Negatives go LEFT**  **Positives go RIGHT**

**A.** 2 + -5

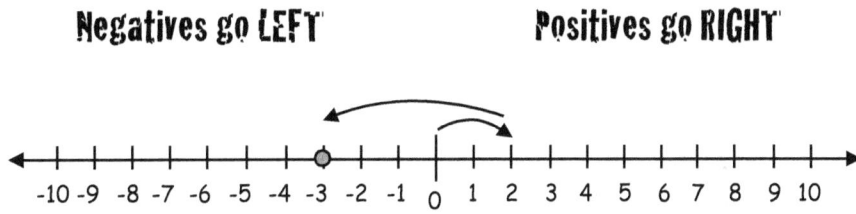

-10 -9 -8 -7 -6 -5 -4 -3 -2 -1 0 1 2 3 4 5 6 7 8 9 10

start at 0

go RIGHT 2   go LEFT 5

---

**Add the integers by drawing arrows on the number line; put a dot on the answer**

**Negatives go LEFT**  **Positives go RIGHT**

**IMPORTANT! Same signs go in the same direction — That's why you ADD them!**

**Opposite signs go in the opposite direction — That's why you SUBTRACT them!**

**1.** 2 + (-3)

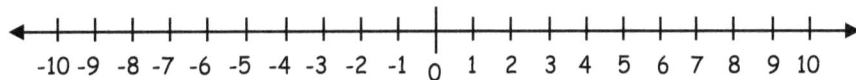

-10 -9 -8 -7 -6 -5 -4 -3 -2 -1 0 1 2 3 4 5 6 7 8 9 10

start at 0

go RIGHT 2    go LEFT 3

> **Notice!** If the negative number is bigger than the positive number you will be left of zero! (negative)

**2.** -4 + (7)

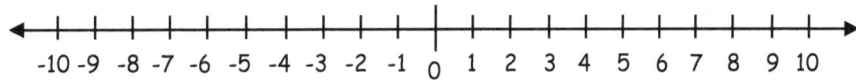

-10 -9 -8 -7 -6 -5 -4 -3 -2 -1 0 1 2 3 4 5 6 7 8 9 10

start at 0

**3.** -1 + (5)

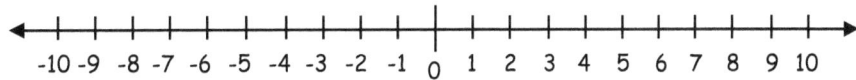

-10 -9 -8 -7 -6 -5 -4 -3 -2 -1 0 1 2 3 4 5 6 7 8 9 10

start at 0

SAME AS + (-6)

**4.** -3 -6

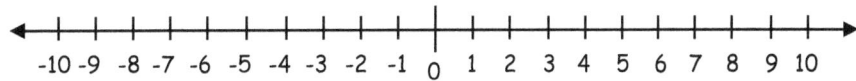

-10 -9 -8 -7 -6 -5 -4 -3 -2 -1 0 1 2 3 4 5 6 7 8 9 10

start at 0

**5.** -8 + 7

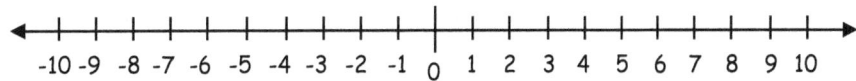

-10 -9 -8 -7 -6 -5 -4 -3 -2 -1 0 1 2 3 4 5 6 7 8 9 10

start at 0

15

# Adding Integers on a Number Line

**Negatives go LEFT**                    **Positives go RIGHT**

**Example**

**A.** -3 - 6

start at 0

go LEFT 3     go LEFT 6 more steps

---

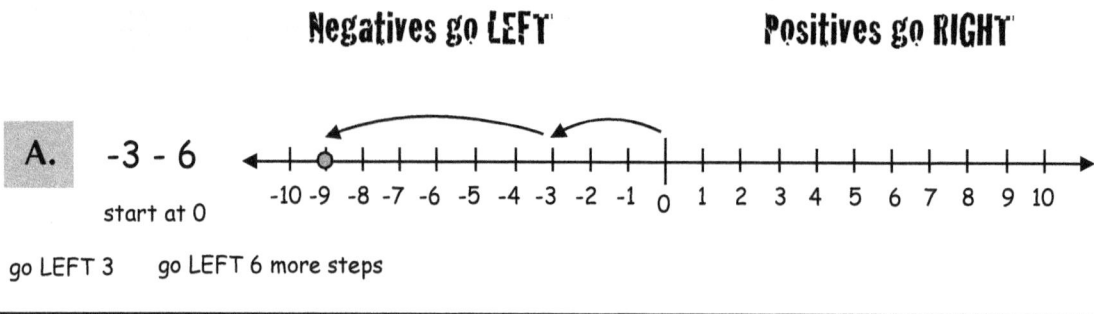

**Add the integers by drawing arrows on the number line; put a dot on the answer**

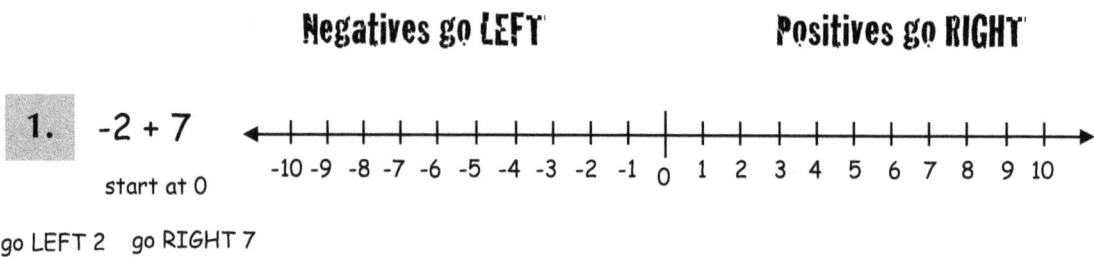

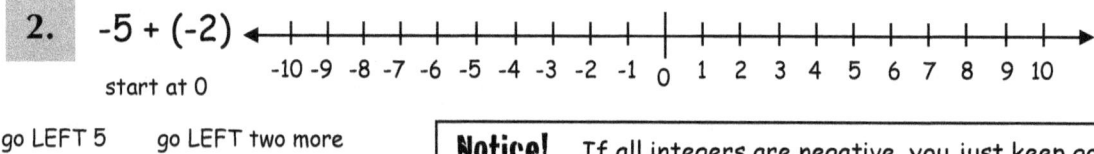

**Negatives go LEFT**                    **Positives go RIGHT**

**1.** -2 + 7

start at 0

go LEFT 2    go RIGHT 7

**2.** -5 + (-2)

start at 0

go LEFT 5      go LEFT two more

**Notice!**   If all integers are negative, you just keep going left

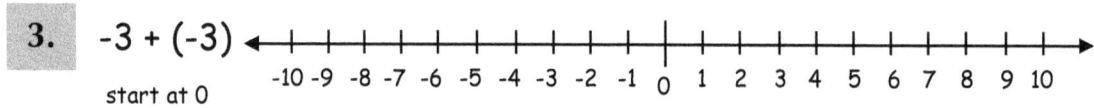

**3.** -3 + (-3)

start at 0

**Notice!**   Your answer is the same distance away from zero as it would be if the answer were POSITIVE

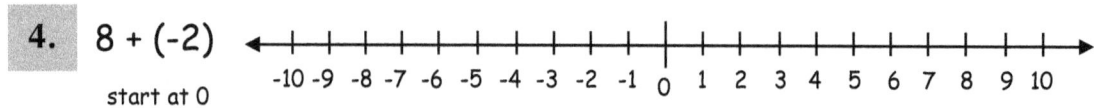

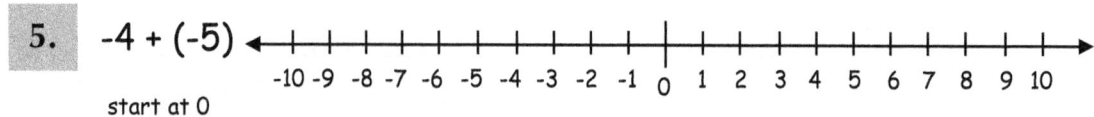

**4.** 8 + (-2)

start at 0

**5.** -4 + (-5)

start at 0

© Peter Wise, 2014

**16**

# Visualizing Integers as Counters

(N) = negative 1    (P) = positive 1

| OPPOSITE SIGN INTEGERS | | SAME SIGN INTEGERS |

**A.**  (-3) + 4 = 1

ONE POSITIVE IS LEFT, SO THE ANSWER IS +1!

N̶ N̶ N̶   P̶ P̶ P̶ P

Cancel out one N for every P
See what is left!

**B.**  (-3) + (-2) = -5

N N N   N N

3 negatives + 2 negatives = five negatives (-5)!

NOTHING TO CANCEL OUT BECAUSE EVERYTHING IS THE SAME SIGN!

## Add the following integers using COUNTERS

**1.**  (-2) + 1 = ☐

N N  P

CANCEL OUT A NEGATIVE AND POSITIVE!

**2.**  (-1) + 2 = ☐

N ○ ○

LABEL THE CIRCLES; THEN CANCEL OUT A NEGATIVE AND A POSITIVE!

**3.**  -1 + (-1) = ☐

○ ○

**4.**  2 + (- 3) = ☐

○ ○ ○
○   ○

**5.**  -2 + 3 = ☐

○ ○ ○
○   ○

**6.**  (-3) + 3 = ☐

○ ○  ○ ○
○   ○

**7.**  (-4) + 3 = ☐

○ ○  ○ ○
○ ○  ○

**8.**  (-4) + (-3) = ☐

○ ○  ○ ○
○ ○  ○

**9.**  (-2) + (-2) + 1 = ☐

○ ○  ○ ○  ○

**10.**  -1 + 3 + (-4) = ☐

○ ○ ○ ○
○   ○ ○

# Enrichment: Substitution with Integers

Use substitution to solve the following problems

**Example**

**A.** $a + b = \boxed{\phantom{xx}}$    $(-2) + (-5) = \boxed{-7}$

$a = -2$    $b = -5$

**3.** $-a + b = (\phantom{xx}) + (\phantom{xx}) = \boxed{\phantom{xx}}$

$a = 10$    $b = -3$

**1.** $a - b = (\phantom{xx}) - (\phantom{xx}) = \boxed{\phantom{xx}}$

$a = 7$    $b = 3$

**4.** $-x + (-y) = (\phantom{xx}) + (\phantom{xx}) = \boxed{\phantom{xx}}$

$x = 3$    $y = 2$

**2.** $-a + b = (\phantom{xx}) + (\phantom{xx}) = \boxed{\phantom{xx}}$

$a = 5$    $b = -6$

**5.** $x + (-y) = (\phantom{xx}) + (\phantom{xx}) = \boxed{\phantom{xx}}$

$x = 6$    $y = 9$

## A LITTLE MORE CHALLENGING . . .

**6.** $a + b + c = (\phantom{xx}) + (\phantom{xx}) + (\phantom{xx}) = \boxed{\phantom{xx}}$

$a = -2$   $b = 10$   $c = -6$

**7.** $x + y - z = \boxed{\phantom{xxxxxxxx}} = \boxed{\phantom{xx}}$

$x = -4$   $y = 20$   $z = 5$

## REPEATED ADDITION OF INTEGERS!

**8.** $b + b = \boxed{\phantom{xx}}$

$b = -4$

**10.** $y + y + y = \boxed{\phantom{xx}}$

$y = -6$

**9.** $a + a + a = \boxed{\phantom{xx}}$

$a = -3$

**11.** $x + x + x + x = \boxed{\phantom{xx}}$

$x = -5$

18

# Elevator-Style Number Lines

Example

**A.** 2 + (-3)

up 2

down -3

START AT ZERO!

POSITIVE GOES UP!

NEGATIVE GOES DOWN!

8
7
6
5
4
3
2
1
0
-1
-2
-3
-4
-5
-6
-7
-8

1. Start at 0
2. Go UP 2
3. Go DOWN 3

---

Draw up and down arrows for the integers; draw a dot on the final answer

Positives go UP          Negatives go DOWN

**1.** 3 + (-4)   **2.** -7 + 8   **3.** -3 + (-3)   **4.** 6 + (-3)   **5.** -2 + (-5)

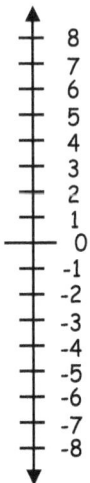
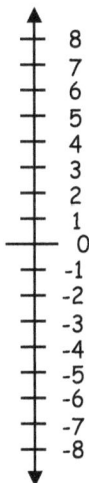
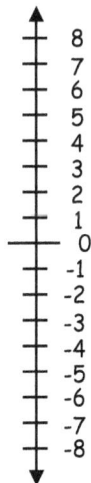
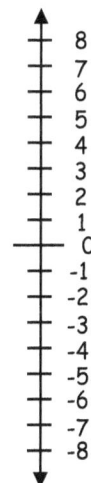
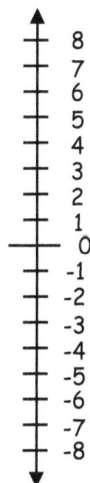

NOTICE! THE ARROWS GO IN **OPPOSITE DIRECTIONS** WHENEVER YOU HAVE ONE POSITIVE AND ONE NEGATIVE INTEGER!

THIS IS WHY YOU **SUBTRACT** THESE INTEGERS!

# Elevator-Style Number Lines

Draw up and down arrows for the integers; draw a dot on the final answer

**Positives go UP**          **Negatives go DOWN**

**1.** 5 + (-6)    **2.** 6 + (-5)    **3.** -3 + (-5)    **4.** -3 + (-4)    **5.** 8 + (-2)

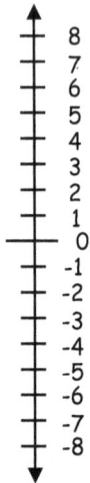 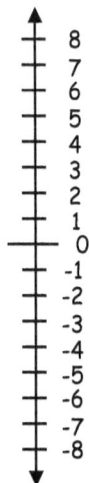 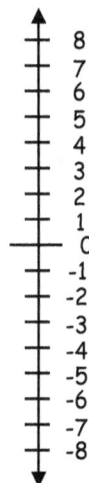 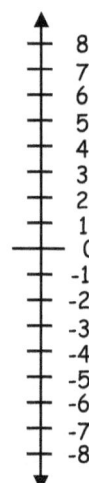 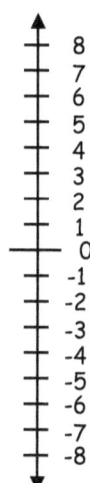

**6.** -4 + 5    **7.** 5 + -4    **8.** -2 + 10    **9.** -2 + (-3)    **10.** 5 + (-7)

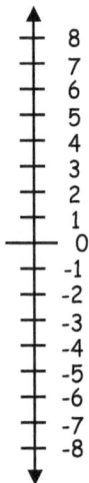 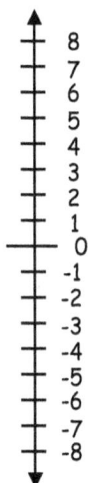 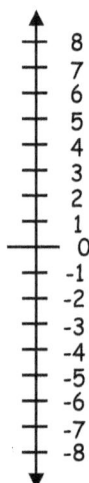 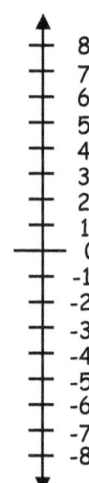 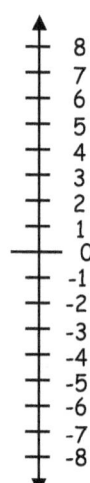

# Adding Groups of Integers

**Example**

Add up all the negatives [ -7 ]

A. -1 + 3 + (-2) + 5 - 4 =

[ -7 + 8 = 1 ]

Add up all the positives [ 8 ]

Now you have just one negative and one positive integer

---

Circle and add up all the negatives [ ]

Add them like they are positive, and then put on a negative sign

1. -2 + 3 + (-4) + 5 - 8 =

[ ]

Add up all the positives [ ]

---

Circle and add up all the negatives [ ]

2. -5 + 3 - 5 + 3 =

[ ]

Add up all the positives [ ]

---

Circle and add up all the negatives [ ]

3. 6 + (-7) + 6 + (-2) =

[ ]

Add up all the positives [ ]

---

Circle and add up all the negatives [ ]

4. 5 + (-50) + 6 + (-50) =

[ ]

Add up all the positives [ ]

© Peter Wise, 2014

21

# Adding Groups of Integers

**1.** -2 + 5 - 2 + 5 = ☐ ← now solve

Negatives = ☐   Positives = ☐

combine all the negatives   combine all the positives

**7.** 4 + (-6) + 14 - 2 = ☐

Negatives = ☐   Positives = ☐

**2.** -4 + 10 + (-4) + 10 = ☐ ←

Negatives = ☐   Positives = ☐

combine all the negatives   combine all the positives

**8.** -12 + 2 - 3 + 7 = ☐

Negatives = ☐   Positives = ☐

**3.** -3 + 20 + (-3) + 20 = ☐

Negatives = ☐   Positives = ☐

**9.** 10 + (-7) + 11 + (-12) = ☐

Negatives = ☐   Positives = ☐

**4.** 4 + (-10) + 4 + (-10) = ☐

Negatives = ☐   Positives = ☐

**10.** -4 + (-5) + (-6) + 7 + 8 = ☐

Negatives = ☐   Positives = ☐

**5.** 7 - 4 + 6 - 5 = ☐

Negatives = ☐   Positives = ☐

**11.** -6 + 8 + 7 + 3 + (-6) = ☐

Negatives = ☐   Positives = ☐

**6.** -12 + 4 + (-7) + 8 = ☐

Negatives = ☐   Positives = ☐

**12.** -12 + 7 - 10 + 6 - 3 + 2 = ☐

Negatives = ☐   Positives = ☐

# Adding and Subtracting Coefficients

Coefficient = number touching (multiplying) a letter

**Examples**

| ADDING VARIABLES | SUBTRACTING VARIABLES |
|---|---|
| **A.** $3x + 4x = 7x$ | **B.** $10a - 2a = 8a$ |

Add or subtract the numbers — Just copy the letter

Add or subtract the numbers — Just copy the letter

1. $2x + 2x = \boxed{\quad x}$

2. $8x - 3x = \boxed{\quad x}$

3. $3a + 10a = \boxed{\quad a}$

4. $12n - 3n = \boxed{\quad}$

5. $2y + 2y + 2y = \boxed{\quad}$

6. $7x + 10x = \boxed{\quad}$

7. $5r + 8r = \boxed{\quad}$

8. $14x - 8x = \boxed{\quad}$

9. $15n - 6n = \boxed{\quad}$

10. $13m - 7m = \boxed{\quad}$

11. $3a - 4a = \boxed{\quad}$

12. $10y + (-12y) = \boxed{\quad}$

13. $-3n + 5n = \boxed{\quad}$

14. $-2y + (-2y) = \boxed{\quad}$

15. $6s + (-7s) = \boxed{\quad}$

16. $-6x + (-6x) = \boxed{\quad}$

17. $-8x + (-3x) = \boxed{\quad}$

18. $16a - 14a = \boxed{\quad}$

19. $-20y + 25y = \boxed{\quad}$

20. $17n - 19n = \boxed{\quad}$

# Concept Quiz

1. If integers have OPPOSITE SIGNS (one positive and one negative) you always ADD / SUBTRACT the numbers before putting on the "winning" sign.

   (circle one)

2. According to the OPPOSITE SIGN RULE, you first pretend that both the positive and negative numbers are:

   _____

3. If the NEGATIVE number is larger than the POSITIVE number, you know that the answer will be POSITIVE / NEGATIVE

   (circle one)

4. On a number line negative numbers go to the LEFT / RIGHT

   (circle one)

5. Add / subtract the following integers by drawing arrows on the number line; put a dot on the answer:

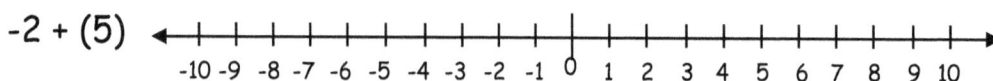

   -2 + (5)

   -10 -9 -8 -7 -6 -5 -4 -3 -2 -1 0 1 2 3 4 5 6 7 8 9 10

6. On a number line, integers with the SAME SIGN always go in THE SAME / OPPOSITE directions

   (circle one)

7. Circle and add up all the negatives [ ]

   -6 + 3 - 6 + 7 =

   Add up all the positives [ ]

   [ ] (put answer here)

# Double Negatives = Positive (or +)

## Double negative = positive

$3 - (- 4) = +7$

THE TWO NEGATIVES TOUCHING BECOME A PLUS!

$-$ touching $-$ equals $+$

**negative** touching **negative** = positive

### Warning!

$- 3 + (- 4) = -7$

THESE NEGATIVES ARE NOT TOUCHING. THEY DO NOT TURN TO A PLUS!

**negative** plus a **negative** = negative

---

## Solve these integer problems; some have double negatives

EVERY TWO NEGATIVES TOUCHING EACH OTHER CANCEL EACH OTHER OUT!

1. $2 - (-4) =$ [ ]

2. $5 - (-6) =$ [ ]

3. $10 - (-3) =$ [ ]

4. $10 - 3 =$ [ ]

5. $- (-12) + 3 =$ [ ]

SAME AS POSITIVE!

6. $16 - (-4) =$ [ ]

7. $9 - (-5) =$ [ ]

8. $-1 - (-2) =$ [ ]

9. $-3 - (-3) =$ [ ]

10. $5 - (-7) =$ [ ]

11. $-2 - (-2) =$ [ ]

12. $-2 - 2 =$ [ ]

13. $- (-3) - (-3) =$ [ ]

14. $(-10) - (-10) =$ [ ]

15. $- (-10) - (-10) =$ [ ]

# Double Negatives = Positive (or +)

**Negative touching negative turns to PLUS**

**Example**

A. $3 - (-2) = 5$

— touching —
negative sign    negative sign

$= +$

one negative sign rotated and joined to the other!

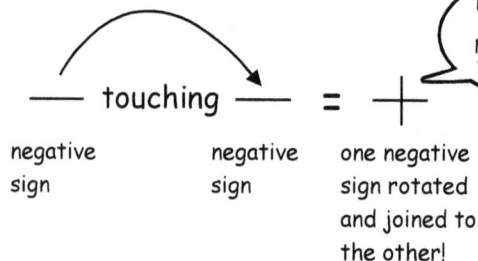

THINK OF A PLUS SIGN AS TWO NEGATIVE SIGNS PUT TOGETHER!

1. $10 - (-2) =$ ☐

2. $10 - 2 =$ ☐

3. $7 - (-5) =$ ☐

4. $7 - 5 =$ ☐

5. $12 - (-4) =$ ☐

6. $14 - (-6) =$ ☐

7. $14 - 6 =$ ☐

8. $-14 - 6 =$ ☐

9. $20 - (-7) =$ ☐

10. $20 - 7 =$ ☐

11. $- (-8) - 4 =$ ☐

12. $-8 - 4 =$ ☐

13. $17 - (-4) =$ ☐

14. $- (-11) - 3 =$ ☐

15. $15 - (-6) =$ ☐

16. $2 - (-7) =$ ☐

17. $- (-5) - (-3) =$ ☐

18. $-6 - (-7) =$ ☐

19. $- 5 - (-5) =$ ☐

20. $-12 - (-10) =$ ☐

# Double Negatives = Positive (or +)

**Example**

Negative touching negative turns to PLUS

A. -2 - (-3) = 1

_SAME AS -2 + 3_

touching = +

negative sign    negative sign    one negative sign rotated and joined to the other!

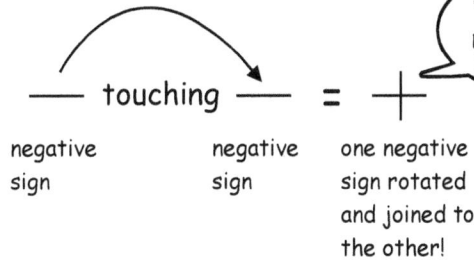

_THINK OF A PLUS SIGN AS TWO NEGATIVE SIGNS PUT TOGETHER!_

1. 8 - (-3) =

2. 8 - 3 =

3. 15 - (-5) =

4. 15 - 5 =

5. 9 - (-3) =

6. 18 - (-4) =

7. 19 - 6 =

8. 14 - (-6) =

9. 22 - (-3) =

10. 22 - 3 =

11. -3 - (-5) =

12. -4 - (-12) =

13. 19 - (-5) =

14. 8 - (-8) =

15. - (-15) - 15 =

16. -2 - (-9) =

17. -3 - (-11) =

18. - (-8) - (-6) =

19. - 5 - (-6) =

20. - (-14) - 12 =

27

# Review: Opposite Sign Integers

Follow the steps; add or subtract the following integers

p = positive    n = negative

1. **Determine the SIGN your answer will be**    2. **Determine the NUMBER**

A. **SIGN** and the
B. **NUMBER**

**Example**

A.    ⓟ n    p ⓝ
   6 + (-7)    answer will be    p ⓝ

7, 6    ← Make both numbers positive.

7 - 6    ← Subtract the smaller number from the larger number.

Answer:
[ - ] [ 1 ]
sign   number

---

**OPPOSITE sign integers always SUBTRACT!**

1.    p n    p n
   10 - 13    answer will be    p n

[ ]    ← Make both numbers positive.

[ ]    ← Subtract the smaller number from the larger number.

Answer:
[ ] [ ]
sign   number

2.    p n    p n
   -7 + (12)    answer will be    p n

[ ]    ← Make both numbers positive.

[ ]    ← Subtract the smaller number from the larger number.

Answer:
[ ] [ ]
sign   number

3.    p n    p n
   37 + (-43)    answer will be    p n

[ ]    ← Make both numbers positive.

[ ]    ← Subtract the smaller number from the larger number.

Answer:
[ ] [ ]
sign   number

4.    p n    p n
   -53 + (49)    answer will be    p n

[ ]    ← Make both numbers positive.

[ ]    ← Subtract the smaller number from the larger number.

Answer:
[ ] [ ]
sign   number

ADDITION RULE FOR NUMBERS WITH OPPOSITE SIGNS:

*PRETEND THEY'RE POSITIVE... SUBTRACT... ADD THE WINNING SIGN!*

# Integer Review

Follow the steps; add or subtract the following integers

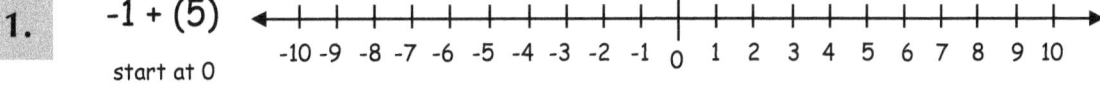

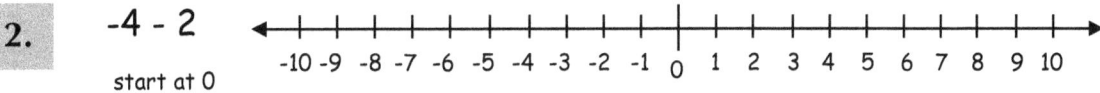

**1.** $-1 + (5)$

start at 0

-10 -9 -8 -7 -6 -5 -4 -3 -2 -1 0 1 2 3 4 5 6 7 8 9 10

**2.** $-4 - 2$

start at 0

-10 -9 -8 -7 -6 -5 -4 -3 -2 -1 0 1 2 3 4 5 6 7 8 9 10

**3.** $(-3) + 1 =$ ☐

LABEL THE CIRCLES WITH EITHER N OR P!

THEN CANCEL OUT ONE NEGATIVE WITH ONE POSITIVE AND SEE WHAT'S LEFT!

**4.** $(-2) + (-4) =$ ☐

MAKE CIRCLES AND PUT "P" FOR POSITIVE AND "N" FOR NEGATIVE; THEN CANCEL TO SEE WHAT'S LEFT!

THE LARGER POSITIVE NUMBER MINUS THE SMALLER POSITIVE NUMBER!

## Opposite Sign Rule

| | Make both numbers positive | | Find the difference (subtract) | Find the sign of the higher number; give your answer this sign | Answer: |
|---|---|---|---|---|---|
| **5.** $-7 + 4 =$ | ☐ | ☐ | ☐ | ☐ | ☐ |
| **6.** $-2 + 10 =$ | ☐ | ☐ | ☐ | ☐ | ☐ |

**7.** $-12 + 8 =$ ☐   **8.** $-3 + 5 =$ ☐   **9.** $-4 + 5 =$ ☐

## Same Sign Rule

**10.** $-2 + (-2) + (-2) =$ ☐      **12.** $-4 + -5 =$ ☐

**11.** $-3 + (-10) + (-10) =$ ☐      **13.** $-2 + -10 =$ ☐

# Integer Addition and Subtraction

| | Circle the correct rule and operation, then solve | | | | |
|---|---|---|---|---|---|

REMEMBER THAT DOUBLE NEGATIVE = POSITIVE!

SAME sign numbers – you always ADD   Just give your answer the same sign as the number you added!

OPPOSITE sign numbers – you always SUBTRACT   Here give your answer the sign of the higher number!

ALWAYS SUBTRACT THE SMALLER NUMBER FROM THE LARGER NUMBER!

| | | | SSR = Same Sign Rule | OSR = Opposite Sign Rule | Will you + or - the numbers? | | ANSWER: |
|---|---|---|---|---|---|---|---|
| 1. | $-3 + (-5)$ | SSR | OSR | add / subtract | = | |
| 2. | $-4 - 2$ | SSR | OSR | add / subtract | = | |
| 3. | $10 + (-6)$ | SSR | OSR | add / subtract | = | |
| 4. | $8 - (-4)$ | SSR | OSR | add / subtract | = | |
| 5. | $-6 - (-2)$ | SSR | OSR | add / subtract | = | |
| 6. | $-18 + 18$ | SSR | OSR | add / subtract | = | |
| 7. | $-10 - (-3)$ | SSR | OSR | add / subtract | = | |
| 8. | $-(-4) + 5$ | SSR | OSR | add / subtract | = | |
| 9. | $-3 + 4$ | SSR | OSR | add / subtract | = | |
| 10. | $-2 - 7$ | SSR | OSR | add / subtract | = | |
| 11. | $15 - 22$ | SSR | OSR | add / subtract | = | |
| 12. | $-4 + (-16)$ | SSR | OSR | add / subtract | = | |

# Integer Addition and Subtraction

| | Circle the correct rule and operation, then solve | | | | |
|---|---|---|---|---|---|

| | | SSR = Same Sign Rule | OSR = Opposite Sign Rule | Will you + or - the numbers? | ANSWER: |
|---|---|---|---|---|---|
| 1. | -5 + 4 | SSR | OSR | add / subtract | = |
| 2. | 4 - 5 | SSR | OSR | add / subtract | = |
| 3. | -6 + (-3) | SSR | OSR | add / subtract | = |
| 4. | -7 + 7 | SSR | OSR | add / subtract | = |
| 5. | -20 + (-1) | SSR | OSR | add / subtract | = |
| 6. | -10 + 2 | SSR | OSR | add / subtract | = |
| 7. | -10 - 10 | SSR | OSR | add / subtract | = |
| 8. | -4 + (-5) | SSR | OSR | add / subtract | = |
| 9. | 10 + (-11) | SSR | OSR | add / subtract | = |
| 10. | 8 - (-2) | SSR | OSR | add / subtract | = |
| 11. | -3 - (-10) | SSR | OSR | add / subtract | = |
| 12. | -(-5) + 3 | SSR | OSR | add / subtract | = |

# Integer Addition and Subtraction

**Add or subtract the following integers**

1. $3 - 4 =$ ☐

2. $-4 + 3 =$ ☐

3. $-3 - 4 =$ ☐

4. $-10 - 6 =$ ☐

5. $-2 + (-3) =$ ☐

6. $7 - (-5) =$ ☐

7. $-9 + (-10) =$ ☐

8. $-6 + (-3) =$ ☐

9. $4 + (-4) =$ ☐

10. $-7 + 8 =$ ☐

11. $8 - 9 =$ ☐

12. $-4 + 10 =$ ☐

13. $-17 + 15 =$ ☐

14. $-21 + 20 =$ ☐

15. Write 15 - 4 as an addition problem with a negative integer ☐

# Integer Addition and Subtraction

**Add or subtract the following integers**

1. $5 - 5 =$ ⬜

2. $5 - 6 =$ ⬜

*WHY IS THIS? (COMPARE THIS PROBLEM TO THE FIRST ONE!)*

3. $7 - 5 =$ ⬜

*COMPARE THESE TWO PROBLEMS!*

4. $5 - 7 =$ ⬜

*WHY IS ONE POSITIVE AND THE OTHER NEGATIVE?*

5. $10 - 2 =$ ⬜

6. $2 - 10 =$ ⬜

*REMEMBER! IF THE LARGER NUMBER IS POSITIVE, THE ANSWER WILL BE POSITIVE!*

7. $6 - 1 =$ ⬜

8. $1 - 6 =$ ⬜

*THESE ARE BOTH NEGATIVE NUMBERS!*

9. $-2 - 4 =$ ⬜

*HOW MANY NEGATIVES DO YOU REALLY HAVE?*

10. $10 - 13 =$ ⬜

11. $-6 + (-6) =$ ⬜

12. $-10 + (-6) =$ ⬜

13. $-10 + 6 =$ ⬜

14. $(-4) - 8 =$ ⬜

15. $-5 + 4 =$ ⬜

16. $12 - 14 =$ ⬜

17. $-7 + (-3) =$ ⬜

18. $-8 + (-8) =$ ⬜

*SAME AS $2 \cdot (-8)$!*

*WHAT HAPPENS WHEN YOU SUBTRACT ONE MORE THAN YOU HAVE?*

19. $10 - 11 =$ ⬜

20. $-5 + (-4) =$ ⬜

21. $20 + (-22) =$ ⬜

22. $-13 + 3 =$ ⬜

23. $-13 + (-3) =$ ⬜

# Integer Addition and Subtraction

| Add or subtract the following integers |
|---|

1.  4 + (-6)  =  ☐

2.  17 + (-5)  =  ☐

3.  -15 - 4  =  ☐

4.  6 - (-4)  =  ☐

5.  (-4) + (-3) =  ☐

6.  (-8) - (-4) =  ☐

7.  18 - 20  =  ☐

8.  2 - (-6)  =  ☐

9.  10 + (-13) =  ☐

10.  (-20) + (-5) =  ☐

11.  -8 - 6  =  ☐

12.  -8 + (-6)  =  ☐

13.  4 - (-7)  =  ☐

14.  -3 - (-5)  =  ☐

15.  12 - (-10) =  ☐

16.  -4 + (-15) =  ☐

17.  -10 + 12  =  ☐

18.  -10 + (-15) =  ☐

# Find the Missing Integers

1. $-2 +$ ☐ $= -4$

2. $-5 -$ ☐ $= -10$

3. $3 +$ ☐ $= 0$

4. $3 +$ ☐ $= -1$

5. $3 +$ ☐ $= -2$

6. $4 +$ ☐ $= 1$

7. $10 +$ ☐ $= -2$

8. ☐ $+ (-4) = 5$

9. ☐ $+ (-11) = -1$

10. $-4 +$ ☐ $= 0$

11. $-4 +$ ☐ $= 6$

12. $8 +$ ☐ $= 0$

13. $8 +$ ☐ $= -2$

14. ☐ $+ (-3) = -9$

15. $6 -$ ☐ $= 10$

16. ☐ $- 3 = -2$

17. $5 +$ ☐ $= -5$

18. ☐ $- (-3) = 9$

# Concept Quiz

1. 5 + (-2) can also be written in what simpler form? _____

2. Integers are whole numbers and their _____ .

Label the following integers with p for positive and n for negative:

3. ☐☐ 10 - 6        4. ☐☐ -7 + (-3)        5. ☐☐ -12 - (-4)

6. When you you **add** integers? _____

7. When do you **subtract** integers? _____

8. What is the "Opposite-Sign Rule?"

    Step #1 _____

    Step #2 _____

    Step #3 _____

9. What is the "Same-Sign Rule?"

    _____

    _____

# Multiplying Integers

**Examples**

ODD number of negative signs in the problem: **NEGATIVE ANSWER**

EVEN number of negative signs in the problem: **POSITIVE ANSWER**

BECAUSE EVERY TWO NEGATIVES CANCEL EACH OTHER OUT!

A. $\boxed{-2 \cdot 3}$ = -6

ODD number of negative signs in the problem: $\boxed{\text{NEGATIVE ANSWER}}$

2 neg's cancel out

B. $\cancel{-}3 \cdot \cancel{-}4$ = 12

EVEN number of negative signs in the problem: $\boxed{\text{POSITIVE ANSWER}}$

---

**Circle EVEN or ODD and POS / NEG; give your product the correct sign**

## Cancel out every two negatives, if none are left, the answer is positive

1. $-5 \cdot 3 =$ ☐

EVEN / ODD number of negative signs in the problem (circle one)

POS / NEG answer (circle one)

5. $8 \cdot (-3) =$ ☐

EVEN / ODD number of negative signs in the problem

POS / NEG answer

2. $-5 \cdot -4 =$ ☐

EVEN / ODD number of negative signs in the problem

POS / NEG answer

WHEN NUMBERS TOUCH, THEY TIMES!

6. $(-6)(-6) =$ ☐

EVEN / ODD number of negative signs in the problem

POS / NEG answer

3. $(-1)(-1)(-1) =$ ☐

EVEN / ODD number of negative signs in the problem

POS / NEG answer

7. $(-1)(-2)(-6) =$ ☐

EVEN / ODD number of negative signs in the problem

POS / NEG answer

4. $(-1)(-1)(-1)(-1) =$ ☐

EVEN / ODD number of negative signs in the problem

POS / NEG answer

8. $(-2)(-2)(-3)(-2) =$ ☐

EVEN / ODD number of negative signs in the problem

POS / NEG answer

**37**

# Multiplying Integers

### Cancel out every two negatives, if none are left, the answer is positive

**1.**  $-5 \cdot -5 =$ ☐

EVEN / ODD
number of negative
signs in the problem
(circle one)

POS / NEG answer
(circle one)

**6.**  $(-8)(-8) =$ ☐

EVEN / ODD
number of negative
signs in the problem

POS / NEG answer

**2.**  $-6 \cdot 3 =$ ☐

EVEN / ODD
number of negative
signs in the problem

POS / NEG answer

**7.**  $(-3)(-3)(3) =$ ☐

EVEN / ODD
number of negative
signs in the problem

POS / NEG answer

**3.**  $(-2)(-2)(-2) =$ ☐

EVEN / ODD
number of negative
signs in the problem

POS / NEG answer

**8.**  $4 \cdot (-7) =$ ☐

EVEN / ODD
number of negative
signs in the problem

POS / NEG answer

**4.**  $(-2)(-2)(-2)(-2) =$ ☐

EVEN / ODD
number of negative
signs in the problem

POS / NEG answer

**9.**  $(-3)(-8)(-1) =$ ☐

EVEN / ODD
number of negative
signs in the problem

POS / NEG answer

**5.**  $(-1)(-2)(-2)(-2) =$ ☐

EVEN / ODD
number of negative
signs in the problem

POS / NEG answer

**10.**  $(-4)(-3)(3) =$ ☐

EVEN / ODD
number of negative
signs in the problem

POS / NEG answer

38

# Multiplying Integers

**Cancel out every two negatives, if none are left, the answer is positive**

1.  $-4 \cdot 8 =$ ☐

EVEN / ODD
number of negative
signs in the problem

POS / NEG answer

6.  $(-7)(4)(-1) =$ ☐

7.  $(-2)(-3)(-5) =$ ☐

2.  $-6 \cdot -8 =$ ☐

EVEN / ODD
number of negative
signs in the problem

POS / NEG answer

8.  $-12 \cdot -2 =$ ☐

9.  $(-7)^2 =$ ☐

3.  $7 \cdot (-2) =$ ☐

EVEN / ODD
number of negative
signs in the problem

POS / NEG answer

10.  $(-3)(-3)(-3) =$ ☐

multiply the numerators

11.  $\dfrac{1}{-2} \cdot \dfrac{1}{3} = \dfrac{\square}{\square\ \square}$

multiply the denominators

4.  $(-1)(-3)(-7) =$ ☐

EVEN / ODD
number of negative
signs in the problem

POS / NEG answer

5.  $-11 \cdot -5 =$ ☐

EVEN / ODD
number of negative
signs in the problem

POS / NEG answer

12.  $(-4)(-2)(7) =$ ☐

# Multiplication as Repeated Addition of Integers

## Multiplication is repeated addition

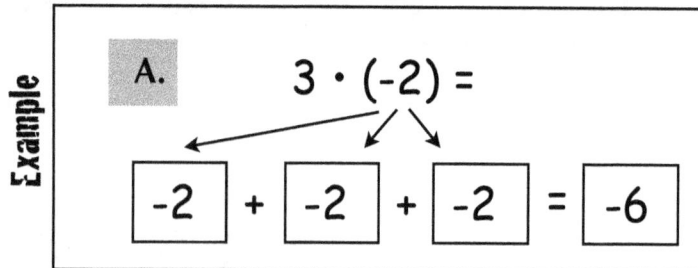

**Example**

**A.** $3 \cdot (-2) =$

| -2 | + | -2 | + | -2 | = | -6 |

---

**Show the multiplication problems as repeated addition; then solve**

**1.** $2 \cdot (-3) =$

$\square + \square = \square$

**5.** $3 \cdot (-4) =$

$\square + \square + \square = \square$

**2.** $4 \cdot (-2) =$

$\square + \square + \square + \square = \square$

**6.** $2 \cdot (-8) =$

$\square + \square = \square$

**3.** $2 \cdot (-5) =$

$\square + \square = \square$

**7.** $4 \cdot (-5) =$

$\square + \square + \square + \square = \square$

Show repeated addition of (-5)

**4.** $2 \cdot (-9) =$

$\square + \square = \square$

**8.** $3 \cdot (-7) =$

$\square + \square + \square = \square$

Show repeated addition of (-7)

# Division as Backwards Multiplication

**Example**

A. $12 \div (-6) = \boxed{-2}$

$= (-6) \cdot \boxed{-2}$

THE ANSWER TO THE BOTTOM PROBLEM TELLS YOU THE ANSWER TO THE TOP PROBLEM!

WORK BACKWARDS!

---

**Solve the following division problems by working BACKWARDS**

1. $18 \div (-3) = \boxed{\phantom{x}}$

   $= (-3) \cdot \boxed{\phantom{x}}$

   same number

   START WITH THE BOTTOM PROBLEM!

7. $-16 \div 2 = \boxed{\phantom{x}}$

   $= 2 \cdot \boxed{\phantom{x}}$

2. $-10 \div (-2) = \boxed{\phantom{x}}$

   $= (-2) \cdot \boxed{\phantom{x}}$

8. $-8 \div (-4) = \boxed{\phantom{x}}$

   $= (-4) \cdot \boxed{\phantom{x}}$

3. $-12 \div 3 = \boxed{\phantom{x}}$

   $= 3 \cdot \boxed{\phantom{x}}$

9. $20 \div (-5) = \boxed{\phantom{x}}$

   $= (-5) \cdot \boxed{\phantom{x}}$

4. $15 \div (-5) = \boxed{\phantom{x}}$

   $= (-5) \cdot \boxed{\phantom{x}}$

10. $-22 \div 2 = \boxed{\phantom{x}}$

    $= 2 \cdot \boxed{\phantom{x}}$

5. $-20 \div 2 = \boxed{\phantom{x}}$

   $= 2 \cdot \boxed{\phantom{x}}$

11. $18 \div (-9) = \boxed{\phantom{x}}$

    $= (-9) \cdot \boxed{\phantom{x}}$

6. $-9 \div (-3) = \boxed{\phantom{x}}$

   $= (-3) \cdot \boxed{\phantom{x}}$

12. $-24 \div (-6) = \boxed{\phantom{x}}$

    $= (-6) \cdot \boxed{\phantom{x}}$

# Integer Division

**Examples**

EVEN number of negative signs in the problem: **POSITIVE ANSWER**

ODD number of negative signs in the problem: **NEGATIVE ANSWER**

2 neg's cancel out

BECAUSE EVERY TWO NEGATIVES CANCEL EACH OTHER OUT!

**A.** $\boxed{-10 \div 2} = -5$

ODD number of negative signs in the problem: $\boxed{\text{NEGATIVE ANSWER}}$

**B.** $-6 \div -2 = 3$

EVEN number of negative signs in the problem: $\boxed{\text{POSITIVE ANSWER}}$

---

**Count the number of negative signs and solve the following division problems**

**1.** $-8 \div (-2) =$ ☐

EVEN / ODD   POS / NEG answer
number of negative
signs in the problem

**2.** $-9 \div (3) =$ ☐

EVEN / ODD   POS / NEG answer
number of negative
signs in the problem

**3.** $-14 \div 7 =$ ☐

EVEN / ODD   POS / NEG answer
number of negative
signs in the problem

**4.** $-32 \div (-8) =$ ☐

EVEN / ODD   POS / NEG answer
number of negative
signs in the problem

**5.** $-27 \div -3 =$ ☐

EVEN / ODD   POS / NEG answer
number of negative
signs in the problem

**7.** $-33 \div 3 =$ ☐

**8.** $-5 \div (-1) =$ ☐

**9.** $42 \div (-6) =$ ☐

**10.** $-36 \div -9 =$ ☐

**11.** $-24 \div (-2) =$ ☐

**12.** $48 \div (-6) =$ ☐

# Substitution with Integers

**1.** $a + b$

$(\quad) + (\quad) = \boxed{\phantom{xx}}$

$a = -2 \quad b = -5$

KEEP THE NEGATIVE YOU SEE HERE; ADD ANOTHER NEGATIVE FROM THE −10!

KEEP ON THE LOOKOUT FOR DOUBLE NEGATIVES!

**3.** $-a + b$

$-(\quad) + (\quad) = \boxed{\phantom{xx}}$

$a = -10 \quad b = 5$

YOU HAVE ONE NEGATIVE SIGN ALREADY, THE −5 ADDS ONE MORE...

...SO YOU REALLY HAVE TWO NEGATIVES IN FRONT OF THE 5!

**2.** $a - b$

$(\quad) - (\quad) = \boxed{\phantom{xx}}$

$a = -2 \quad b = -5$

**4.** $-x + (-y)$

$(\quad) - (\quad) = \boxed{\phantom{xx}}$

$x = 9 \quad y = -4$

## A LITTLE MORE CHALLENGING ...

WHEN THEY *TOUCH*, THEY *TIMES!*

**5.** $x + y - z$

$(\quad) + (\quad) - (\quad) = \boxed{\phantom{xx}}$

$x = -7 \quad y = -3 \quad z = -12$

**6.** $(a)(b + c)$

$(\quad)(\quad + \quad) = \boxed{\phantom{xx}}$

$a = -4 \quad b = -2 \quad c = -5$

## INTEGER MULTIPLICATION REVIEW

**7.** $(-1)(-1) = \boxed{\phantom{xx}}$

**10.** $8 \cdot 6 \cdot (-1) = \boxed{\phantom{xx}}$

**8.** $(-1)(-1)(-1) = \boxed{\phantom{xx}}$

**11.** $(-1)(7)(-9)(-1) = \boxed{\phantom{xx}}$

**9.** $(-1)(-1)(-1)(-1) = \boxed{\phantom{xx}}$

**12.** $(-1)^5 = \boxed{\phantom{xx}}$

5 NEGATIVE ONES ARE MULTIPLYING EACH OTHER!

# Addition/Subtraction Review

| Add or subtract the following integers |
|---|

**SSR** = Same Sign Rule    **OSR** = Opposite Sign Rule

Just give your answer the same sign as the number you added!

Here give your answer the sign of the higher number!

**SAME** sign numbers – you always **ADD**

**OPPOSITE** sign numbers – you always **SUBTRACT**

SSR          OSR

1. $-1 + (-2) =$ [ ]    SSR    OSR

2. $-3 - 4 \quad =$ [ ]    SSR    OSR

3. $-2 + 7 \quad =$ [ ]    SSR    OSR

4. $7 - 2 \quad =$ [ ]    SSR    OSR

5. $5 - 6 \quad =$ [ ]    SSR    OSR

6. $-8 - 9 \quad =$ [ ]    SSR    OSR

7. $-2 - 4 - 2 =$ [ ]    SSR    OSR

8. $-10 + (-10) + (-10) =$ [ ]    SSR    OSR

9. $-3 - 3 + 3 + 3 \quad =$ [ ]    SSR    OSR    BOTH

10. $- -6 \quad =$ [ ]

> HINT: EVERY TWO NEGATIVES CANCEL EACH OTHER OUT!

11. $- - -6 \quad =$ [ ]

> SHORTCUT! AN **EVEN** NUMBER OF NEGATIVES = **POSITIVE** NUMBER, **ODD** NUMBER OF NEGATIVES = **NEGATIVE** NUMBER

12. $- - - -6 =$ [ ]

44

# Addition/Subtraction Review

| Add or subtract the following integers |
|---|

1. 5 - 6 = ☐

2. -5 - 6 = ☐

3. -10 - 4 = ☐

4. -10 + 4 = ☐

5. -10 - 10 = ☐

6. -10 + 10 = ☐

7. 16 + (-17) = ☐

8. 4 + (-8) = ☐

9. -30 + 20 = ☐

10. -16 + (-16) = ☐

11. -2 - 8 = ☐

12. 7 + (-9) = ☐

13. 6 + (-3) = ☐

14. -10 + 12 = ☐

15. 9 + (-10) = ☐

16. -9 + (-10) = ☐

17. -100 + 4 = ☐

18. -100 - 4 = ☐

# Integers: All Operations

1. -3 + (-3) = ☐

2. 3 · (-3) = ☐

3. 8 - 12 = ☐

4. -4 · (3) = ☐

5. -16 - 4 = ☐

6. -35 ÷ 5 = ☐

7. -7 · 5 = ☐

8. 12 ÷ (-3) = ☐

9. -6 + (-3) = ☐

10. -28 ÷ (-4) = ☐

11. 9 · (-3) = ☐

12. 12 + (-2) = ☐

13. 4 - 10 = ☐

14. -7 + (17) = ☐

15. 6 · (-5) = ☐

16. -80 ÷ 8 = ☐

17. -3 + 7 = ☐

18. 100 ÷ (-50) = ☐

# Integers: All Operations

1. $-7 \cdot 3 = \boxed{\phantom{xxx}}$

2. $-7 \cdot (-3) = \boxed{\phantom{xxx}}$

3. $(-9) + (-9) = \boxed{\phantom{xxx}}$

4. $(8)(-4) = \boxed{\phantom{xxx}}$

WHEN THEY TOUCH THEY TIMES!

5. $-16 \div 8 = \boxed{\phantom{xxx}}$

6. $14 - 5 = \boxed{\phantom{xxx}}$

7. $-6 \cdot 7 = \boxed{\phantom{xxx}}$

8. $18 - 20 = \boxed{\phantom{xxx}}$

9. $-20 \div (-4) = \boxed{\phantom{xxx}}$

10. $5 - (-6) = \boxed{\phantom{xxx}}$

11. $-5 - (-6) = \boxed{\phantom{xxx}}$

12. $40 \div (-4) = \boxed{\phantom{xxx}}$

13. $15 + (-3) = \boxed{\phantom{xxx}}$

14. $15 \div (-3) = \boxed{\phantom{xxx}}$

15. $(-7)(-7) = \boxed{\phantom{xxx}}$

16. $64 \div (-8) = \boxed{\phantom{xxx}}$

17. $-6 \cdot (-3) = \boxed{\phantom{xxx}}$

18. $(-2)(-3)(-4) = \boxed{\phantom{xxx}}$

# Fact Families with Negative Numbers

**Example**

|  | | MULTIPLICATION | DIVISION |
|---|---|---|---|
| **A.** | (-3, 4, -12) | -3 · 4 = -12 | -12 ÷ -3 = 4 |
|  | | 4 · (-3) = -12 | -12 ÷ 4 = -3 |

switched   same   same   switched

---

## Give the four fact families for each set of integers

**1.** (3, -8, -24)

multiplication

division

**2.** (-6, 12, -72)

multiplication

division

**3.** (-7, -8, 56)

multiplication

division

**4.** (6, -9, -54)

multiplication

division

# Concept Quiz

1. If you have an EVEN number of negative signs in a multiplication or division problem the answer will be positive/negative. (circle the correct answer)

2. Why does this happen? _____

   _____

3. If you have an ODD number of negative signs in a multiplication or division problem the answer will be positive/negative. (circle the correct answer)

4. One way to think of division problems is to view them as _____ multiplication.

5. Fill in the following chart for the fact family for   (-28, 7, -4)

<table>
<tr><td style="text-align:center">multiplication</td><td style="text-align:center">division</td></tr>
<tr><td></td><td></td></tr>
<tr><td></td><td></td></tr>
</table>

6. You can think of multiplication of integers as _____ addition.

7. Write this as an addition problem: $2 \cdot (-8)$ _____

8. BONUS: Calculate $(-3)^3$ _____

# Basic Equalities

**Example**

A. $x = -7$ — Put a solid circle on -7
Label it x

● ← a number that equals x (or another letter)

$y = 2$ — Put a solid circle on 2
Label it y

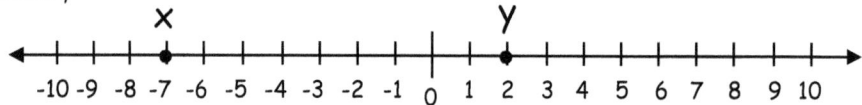

x        y

-10 -9 -8 -7 -6 -5 -4 -3 -2 -1 0 1 2 3 4 5 6 7 8 9 10

DOT means EQUALS

---

## Put solid circles (dots) on the number line and label them

**1.** $x = 5$
$y = -2$

-10 -9 -8 -7 -6 -5 -4 -3 -2 -1 0 1 2 3 4 5 6 7 8 9 10

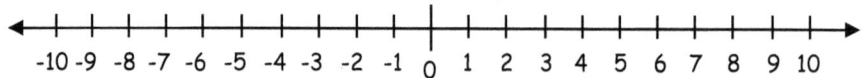

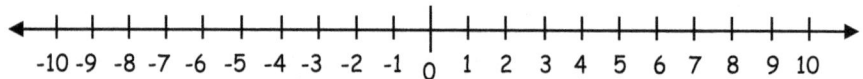

How far is x from zero? ☐   Left? Right?   How far is y from zero? ☐   Left? Right?

(circle one)

NOTE: DISTANCE IS ALWAYS POSITIVE!

What is the distance between the two points? ☐

**2.** $x = -8$
$y = -4$

-10 -9 -8 -7 -6 -5 -4 -3 -2 -1 0 1 2 3 4 5 6 7 8 9 10

How far is x from zero? ☐   Left? Right?   How far is y from zero? ☐   Left? Right?

(circle one)

What is the distance between the two points? ☐

**3.** $a = -10$
$b = -8$

-10 -9 -8 -7 -6 -5 -4 -3 -2 -1 0 1 2 3 4 5 6 7 8 9 10

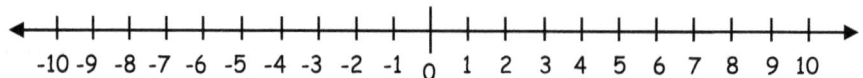

How far is a from zero? ☐   Left? Right?   How far is b from zero? ☐   Left? Right?

What is the distance between the two points? ☐

**4.** $r = -6$
$s = 1$

-10 -9 -8 -7 -6 -5 -4 -3 -2 -1 0 1 2 3 4 5 6 7 8 9 10

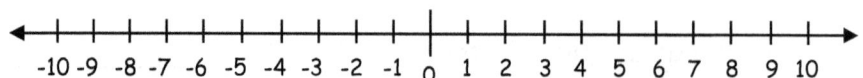

How far is r from zero? ☐   Left? Right?   How far is s from zero? ☐   Left? Right?

What is the distance between the two points? ☐

**50**

# Basic Inequalities

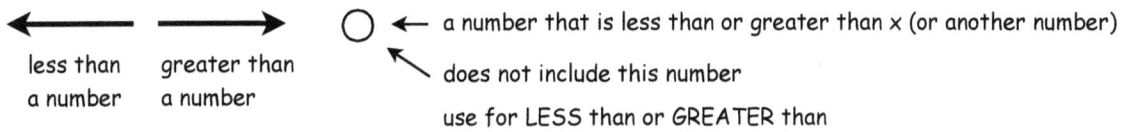

**Example**

less than a number ← →  greater than a number

○ ← a number that is less than or greater than x (or another number)

↖ does not include this number

use for LESS than or GREATER than

**A.** x < 4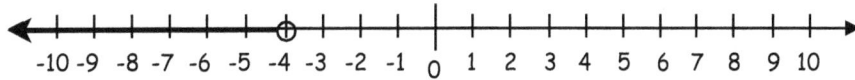

x is LESS than 4

**1.** Put a hollow circle on 4     **2.** Draw an arrow going left     LESS = LEFT

---

**Add the integers by drawing arrows on the number line; put hollow circles where they go**

## Less than = arrow goes LEFT          Greater than = arrow goes RIGHT

**1.** x > 2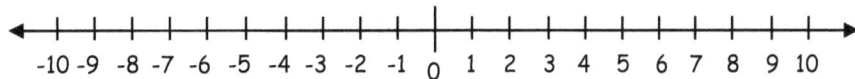

**1.** Put a hollow circle on 2     **2.** Draw an arrow going right     GREATER = RIGHT

**2.** x < -1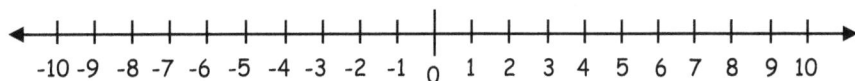

**3.** x > -6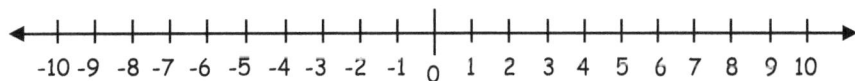

**4.** x < 5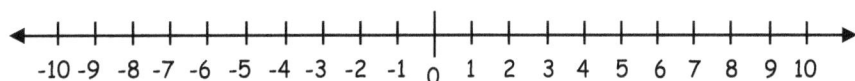
start at 0

**5.** x > -4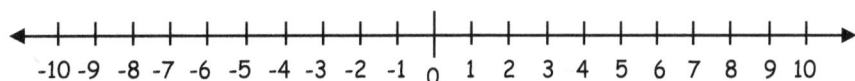

# Greater Than OR Equal To

Example

← less than a number    greater than a number →

● ← a number that equals x (or another letter)

**A.** x ≤ -3

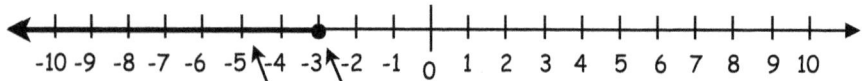

-10 -9 -8 -7 -6 -5 -4 -3 -2 -1 0 1 2 3 4 5 6 7 8 9 10

BOTH ARE COMBINED IN THIS SIGN!

x = -3 and x ≤ -3

"LESS THAN OR EQUAL TO"

less than 3

OR equals (includes) 3

x ≤ -3

**Memory Trick**

**a.** Put a hollow circle ("donut") if the point on the number line "Do-not" equal a number

**b.** Put a solid circle ("dot") if it "Do" equal a number

---

## Write inequalities shown by the number line

**1.** What inequality does this line show?     x ≤ ☐

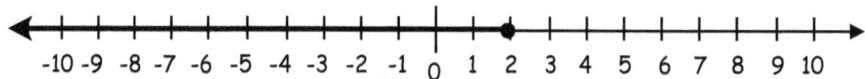

-10 -9 -8 -7 -6 -5 -4 -3 -2 -1 0 1 2 3 4 5 6 7 8 9 10

**2.** What inequality does this line show?     ☐

-10 -9 -8 -7 -6 -5 -4 -3 -2 -1 0 1 2 3 4 5 6 7 8 9 10

**3.** What inequality does this line show?     ☐

-10 -9 -8 -7 -6 -5 -4 -3 -2 -1 0 1 2 3 4 5 6 7 8 9 10

**4.** What inequality does this line show?     ☐

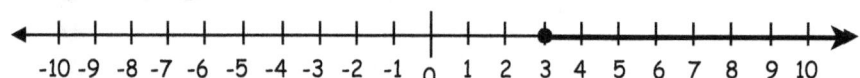

-10 -9 -8 -7 -6 -5 -4 -3 -2 -1 0 1 2 3 4 5 6 7 8 9 10

# Greater Than OR Equal To

## Which inequalities do these number lines show?

**1.**

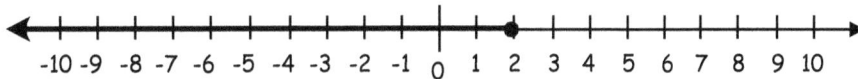

Circle the correct answer:  (a) x < 2  (b) x = 2  (c) x ≤ 2

**2.**

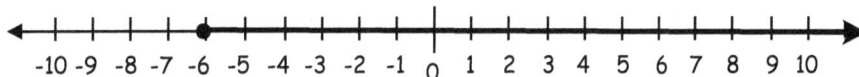

Circle the correct answer:  (a) x ≥ -6  (b) x = -6  (c) x < -6

**3.**

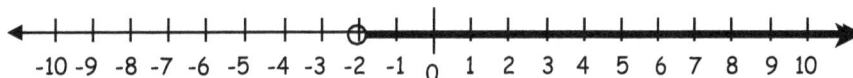

Circle the correct answer:  (a) x ≥ -2  (b) x > -2  (c) x < -2

**4.**

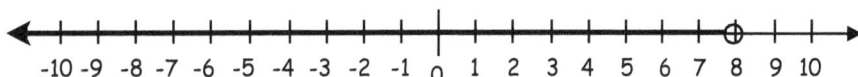

Circle the correct answer:  (a) x ≤ 8  (b) x ≥ 8  (c) x < 8

## Draw the inequalities

**5.** Show x ≥ -5 on the number line:

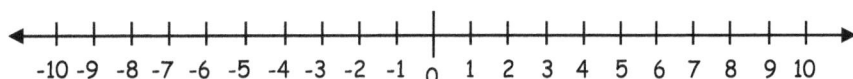

**6.** x > 5

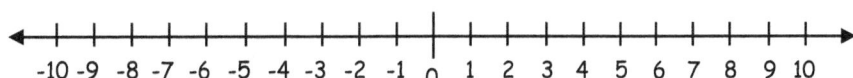

# Mixed Inequalities

**Write inequalities shown by each number line**

1. What inequality does this line show?

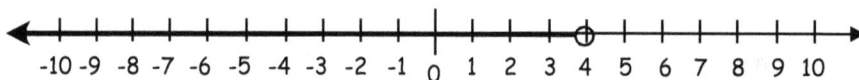

2. What inequality does this line show?

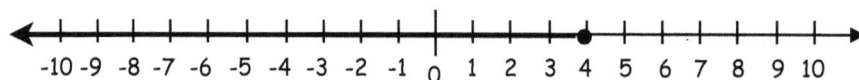

3. What inequality does this line show?

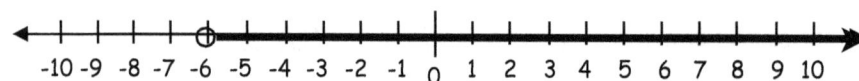

4. What inequality does this line show?

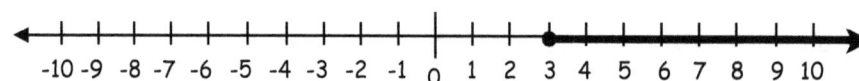

5. What inequality does this line show?

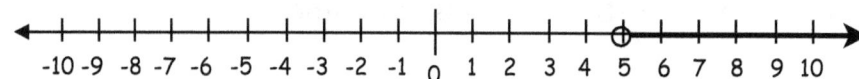

6. What inequality does this line show?

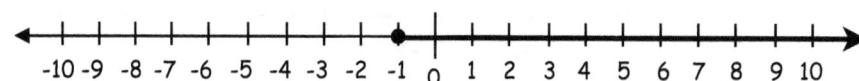

# Mixed Inequalities

**1.** What inequality does this line show?

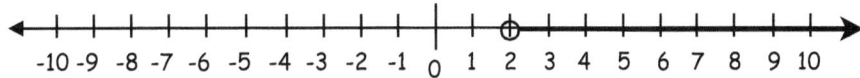

```
  +---+---+---+---+---+---+---+---+---+---|---+---⊕---+---+---+---+---+---+---+---+--->
 -10 -9 -8 -7 -6 -5 -4 -3 -2 -1  0  1  2  3  4  5  6  7  8  9  10
```

**2.** What inequality does this line show?

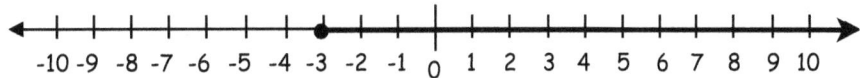

```
  +---+---+---+---+---+---+---●---+---+---|---+---+---+---+---+---+---+---+---+---+--->
 -10 -9 -8 -7 -6 -5 -4 -3 -2 -1  0  1  2  3  4  5  6  7  8  9  10
```

**3.** What inequality does this line show?

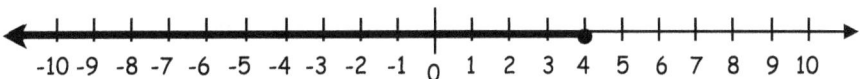

```
◄━━━━━━━━━━━━━━━━━━━━━━━━━━━━━━━━━━━━━━●---+---+---+---+---+---+--->
 -10 -9 -8 -7 -6 -5 -4 -3 -2 -1  0  1  2  3  4  5  6  7  8  9  10
```

**4.** What inequality does this line show?

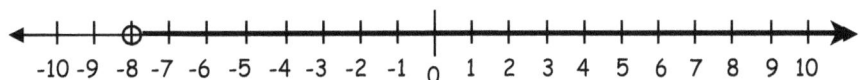

```
  +---+---⊕━━━━━━━━━━━━━━━━━━━━━━━━━━━━━━━━━━━━━━━━━━━━━━━━━━━━━►
 -10 -9 -8 -7 -6 -5 -4 -3 -2 -1  0  1  2  3  4  5  6  7  8  9  10
```

**5.** What inequality does this line show?

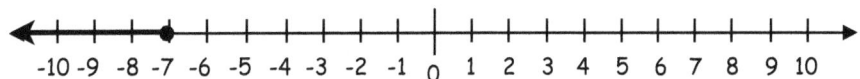

```
◄━━━━━━●---+---+---+---+---+---|---+---+---+---+---+---+---+---+--->
 -10 -9 -8 -7 -6 -5 -4 -3 -2 -1  0  1  2  3  4  5  6  7  8  9  10
```

**6.** What inequality does this line show?

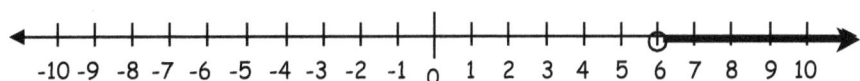

```
  +---+---+---+---+---+---+---+---+---+---|---+---+---+---+---⊕━━━━━━━━━━━━►
 -10 -9 -8 -7 -6 -5 -4 -3 -2 -1  0  1  2  3  4  5  6  7  8  9  10
```

55

# Mixed Inequalities

**Show the following inequalities on the number line**

1. Show x < 6 on the number line:

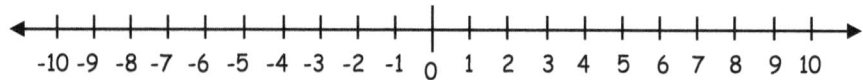

```
  -10 -9 -8 -7 -6 -5 -4 -3 -2 -1  0  1  2  3  4  5  6  7  8  9 10
```

2. Show x ≥ 7 on the number line:

*REMEMBER THAT ON A NUMBER LINE THE = BECOMES A SOLID DOT AND THE > OR < BECOMES AN ARROW!*

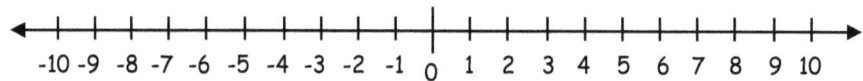

```
  -10 -9 -8 -7 -6 -5 -4 -3 -2 -1  0  1  2  3  4  5  6  7  8  9 10
```

3. Show x ≤ -5 on the number line:

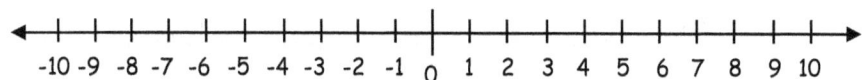

```
  -10 -9 -8 -7 -6 -5 -4 -3 -2 -1  0  1  2  3  4  5  6  7  8  9 10
```

4. Show x > 1 on the number line:

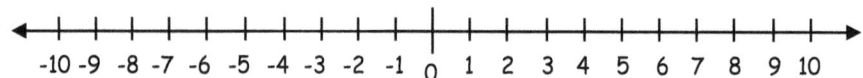

```
  -10 -9 -8 -7 -6 -5 -4 -3 -2 -1  0  1  2  3  4  5  6  7  8  9 10
```

5. Show x < -3 on the number line:

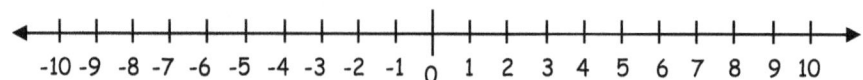

```
  -10 -9 -8 -7 -6 -5 -4 -3 -2 -1  0  1  2  3  4  5  6  7  8  9 10
```

6. Show x ≥ -6 on the number line:

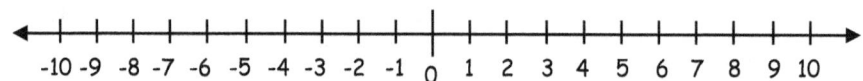

```
  -10 -9 -8 -7 -6 -5 -4 -3 -2 -1  0  1  2  3  4  5  6  7  8  9 10
```

# Multiplication with Fraction Negatives

## Multiply the following fractions

### Examples

How to multiply fractions: multiply the top, multiply the bottom!

**A.** $\dfrac{1}{2} \cdot \dfrac{1}{-6} = \dfrac{\boxed{1 \cdot 1}}{\boxed{2 \cdot (-6)}}$ or $- \dfrac{\boxed{1}}{\boxed{12}}$

**B.** $\dfrac{2}{-3} \cdot \dfrac{1}{-5} = \dfrac{\boxed{2}}{\boxed{15}}$

These two negatives cancel out

---

**1.** $\dfrac{2}{3} \cdot \dfrac{1}{-7} = \square \dfrac{\boxed{\phantom{x}}}{\boxed{\phantom{x}}}$

USE THIS BOX WHENEVER THE FRACTION IS NEGATIVE!

### Count the number of negative signs in both fractions (numerators and denominators)

**A.** If you have an EVEN number of neg. signs, the fraction is positive, because every two negatives cancel each other out!

**B.** If you have an ODD number of neg. signs, the fraction is negative, because after canceling every two negative signs, you still have one left!

**2.** $\dfrac{-5}{6} \cdot \dfrac{1}{3} = \square \dfrac{\boxed{\phantom{x}}}{\boxed{\phantom{x}}}$

**3.** $\dfrac{-2}{9} \cdot \dfrac{-1}{5} = \square \dfrac{\boxed{\phantom{x}}}{\boxed{\phantom{x}}}$

**7.** $\dfrac{-2}{-3} \cdot \dfrac{-4}{-5} = \square \dfrac{\boxed{\phantom{x}}}{\boxed{\phantom{x}}}$

**4.** $\dfrac{-3}{5} \cdot \dfrac{2}{7} = \square \dfrac{\boxed{\phantom{x}}}{\boxed{\phantom{x}}}$

**8.** $\dfrac{-1}{-2} \cdot \dfrac{-3}{4} = \square \dfrac{\boxed{\phantom{x}}}{\boxed{\phantom{x}}}$

**5.** $\dfrac{2}{7} \cdot \dfrac{4}{-9} = \square \dfrac{\boxed{\phantom{x}}}{\boxed{\phantom{x}}}$

**9.** $\dfrac{-2}{11} \cdot \dfrac{3}{-5} = \square \dfrac{\boxed{\phantom{x}}}{\boxed{\phantom{x}}}$

**10.** $\dfrac{-3}{-1} \cdot \dfrac{-5}{7} = \square \dfrac{\boxed{\phantom{x}}}{\boxed{\phantom{x}}}$

**6.** $\dfrac{7}{-8} \cdot \dfrac{1}{-3} = \square \dfrac{\boxed{\phantom{x}}}{\boxed{\phantom{x}}}$

# Negative Numbers with Exponents

**Examples**

## IF THE BASE NUMBER IS NEGATIVE. . .

**Even exponent = positive answer**

*IF THE EXPONENT IS AN EVEN NUMBER THE RESULT IS ALWAYS POSITIVE!*

$-5^2$

| -5 | × | -5 | = | 25 |

*THESE TWO NEGATIVES TURN TO A POSITIVE!*

**ODD exponent = negative answer**

*IF THE EXPONENT IS AN ODD NUMBER THEN THE QUOTIENT WILL BE NEGATIVE!*

*...BECAUSE AFTER EVERY TWO NEGATIVE SIGNS CANCEL OUT ONE WILL STILL BE LEFT OVER!*

$-2^3$

| -2 | × | -2 | × | -2 | = | |

With multiplication and division every two negatives signs cancel each other out!

One negative factor is left over...

...so the answer is negative!

## Strange but true...

-5 is the same as (-1)(5) ...so -5 × -5 can also be written:

*A NEGATIVE SIGN IS THE SAME AS (-1) MULTIPLYING THE POSITIVE NUMBER!*

$-a = (-1)a$

| (-1)(5) | × | (-1)(5) |

With multiplication order doesn't matter so you can rearrange the factors!

| (-1)(-1) | × | (5)(5) |

(+1) × (25)

Every factor is positive, so the answer is positive!

## Multiply the following negative numbers with exponents

Hint: Just multiply as though the base numbers were positive and then decide the sign!

1. $-1^2 = \boxed{\phantom{00}}$

2. $-1^3 = \boxed{\phantom{00}}$

3. $-2^2 = \boxed{\phantom{00}}$

4. $-10^2 = \boxed{\phantom{00}}$

5. $-10^3 = \boxed{\phantom{00}}$

6. $-2^3 = \boxed{\phantom{00}}$

7. $-6^1 = \boxed{\phantom{00}}$

8. $-7^2 = \boxed{\phantom{00}}$

9. $-1^{15} = \boxed{\phantom{00}}$

# Intro to Absolute Value

**Examples**

Absolute value is always POSITIVE (or zero); NEVER NEGATIVE

Absolute value is the DISTANCE from ZERO (which of course can never be negative)

The absolute value of a number is represented by two parallel, vertical lines $|3|$
↑ ↑

3 AWAY FROM ZERO!
$|3| = 3$

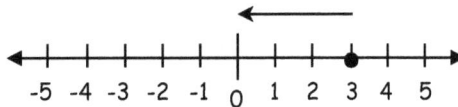

-5 -4 -3 -2 -1 0 1 2 3 4 5

−3 IS ALSO 3 UNITS AWAY FROM ZERO!
$|-3| = 3$ also!

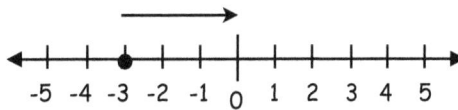

-5 -4 -3 -2 -1 0 1 2 3 4 5

BOTH +3 AND −3 ARE THREE AWAY FROM ZERO!

---

## Answer the following absolute value problems

**1.** $|-7| = $ ☐

Just make the value inside positive

**3.** $|-2| = $ ☐

**5.** $|9| = $ ☐

**2.** $|8| = $ ☐

Just keep the value positive

**4.** $|-a| = $ ☐

**6.** $|-12| = $ ☐

---

### A LITTLE HARDER...

**7.** $|1-5| = $

$|\ ☐\ | = ☐$

SOLVE THE PROBLEM INSDE FIRST, THEN MAKE YOUR ANSWER POSITIVE!

**8.** $|-2 -7| = $

$|\ ☐\ | = ☐$

**9.** $|10 + (-13)| = $

$|\ ☐\ | = ☐$

**10.** $|(-3)(4)| = $

$|\ ☐\ | = ☐$

**11.** $|(-2)(-2)(-5)| = $

$|\ ☐\ | = ☐$

© Peter Wise, 2014

59

# Absolute Value and the Opposite Sign Rule

Aha! Now that you know about absolute value you will learn some interesting things about integers! Think about the sign rules!

**Normal Subtraction**   **Absolute Value**

$5 - 2 =$

1a. ☐

$|5 - 2| =$

1b. $|\ \square\ | = \square$

$2 - 5 =$

2a. ☐

$|2 - 5| =$

2b. $|\ \square\ | = \square$

WHAT DO YOU NOTICE?

---

$10 - 1 =$ ☐   $|10 - 1| =$

3b. $|\ \square\ | = \square$

$1 - 10 =$ ☐   $|1 - 10| =$

4b. $|\ \square\ | = \square$

---

$4 + (-3) =$

5a. ☐

$|4 + (-3)| =$

5b. $|\ \square\ | = \square$

$-3 + 4 =$

6a. ☐

$|(-3) + 4| =$

6b. $|\ \square\ | = \square$

---

OBSERVATIONS:    What is the relationship between a - b and b - a?

_____

_____

---

### Practice with the following absolute value problems

$|2-6| =$

7. $|\ \square\ | = \square$

$|8 + (-10)| =$

8. $|\ \square\ | = \square$

$|7 + (-4)| =$

9. $|\ \square\ | = \square$

# Practice with Absolute Value

Absolute value

- Measures distance from zero
- Always positive (or zero); never negative

## Use your knowledge of ABSOLUTE VALUE to solve the following problems

1. $|-5| = \boxed{\phantom{00}}$

2. $|-7|$ is $\boxed{\phantom{00}}$ units away from zero

3. $|-12| + |-4| = \boxed{\phantom{00}}$

4. $|-10| - |-2| = \boxed{\phantom{00}}$

5. $|-a| = \boxed{\phantom{00}}$

6. $|-x| + |-x| = \boxed{\phantom{00}}$

7. $|-4| \times |-6| = \boxed{\phantom{00}}$

8. $|-20| \times |-2| = \boxed{\phantom{00}}$

9. $\left|-\dfrac{2}{3}\right| = \boxed{\phantom{00}}$

10. $|-15| \times |-10| = \boxed{\phantom{00}}$

## Order these values from LEAST to GREATEST

11. $|-12| \quad |-4| \quad |-1|$

$\boxed{\phantom{xxxxxxxxxxxxxx}}$

LEAVE YOUR ANSWERS IN THE SAME ABSOLUTE VALUE AS ABOVE!

12. $|-6| \quad 5 \quad |-8|$

$\boxed{\phantom{xxxxxxxxxxxxxx}}$

13. $|-10| \quad |-2| \quad 18$

$\boxed{\phantom{xxxxxxxxxxxxxx}}$

14. $\left|-\dfrac{1}{4}\right| \quad \left|-\dfrac{1}{5}\right|$

$\boxed{\phantom{xxxxxxxxxxxxxx}}$

# Absolute Value and the Same Sign Rule

Aha! Now that you know about absolute value you will learn some interesting things about integers! Think about the Sign Rules!

## Normal Addition/Subtraction    Absolute Value

-2 - 2 =

**1a.** ☐

|-2 - 2| =

**1b.** |☐| = ☐

4 + 4 =

**2a.** ☐

|4 + 4| =

**2b.** |☐| = ☐

WHAT DO YOU NOTICE?

-6 - 6 =

**3a.** ☐

|-6 - 6| =

**3b.** |☐| = ☐

-7 + (-5) =

**5a.** ☐

|-7 + (-5)| =

**5b.** |☐| = ☐

-3 + (-5) =

**4a.** ☐

|-3 + (-5)| =

**4b.** |☐| = ☐

-4 - 6 =

**6a.** ☐

|-4 - 6)| =

**6b.** |☐| = ☐

OBSERVATIONS:    What is the relationship between a + b and -a + (-b)?

_____

_____

---

### Practice with the following absolute value problems

|6 + (-8)| =

**7.** |☐| = ☐

|-7 - 7| =

**8.** |☐| = ☐

|15 + (-3)| =

**9.** |☐| = ☐

**62**

# Integer Test

**1.** $4 - 10 =$ ☐

**2.** $-8 - (-4) =$ ☐

**3.** $-12 - 7 =$ ☐

**4.** $5 + (-7) + 8 + (-3) =$ ☐

**5.** $-13 + 9 + (-8) + 5 + (-6) =$ ☐

**6.** $7 \cdot (-9) =$ ☐

**7.** $(-2)(-2)(-2) =$ ☐

**8.** $-24 \div (-4) =$ ☐

**9.** $35 \div (-7) =$ ☐

**10.** $6x + (-14x) =$ ☐

**11.** $-9x + (-7x) =$ ☐

**12.** $-(-7) - 6 =$ ☐

**13.** $7 +$ ☐ $= -7$

**14.** ☐ $- 8 = -20$

**15.** $-x + (-y) =$ ☐

$x = 12 \qquad y = -5$

**16.** $-9^2 =$ ☐

**17.** $|-8| =$ ☐

**18.** $|(-4)(7)| =$ ☐

**19.** Order these from LEAST to GREATEST

$|-14| \quad |-6| \quad |-2|$

☐

**20.** Show $x < -3$ on the number line:

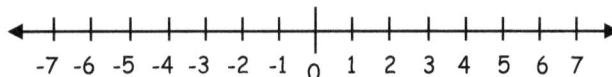

$$\longleftarrow \mid \ \mid \ \mid \ \mid \ \mid \ \mid \ \mid \ \mid \ \mid \ \mid \ \mid \ \mid \ \mid \ \mid \longrightarrow$$
-7 -6 -5 -4 -3 -2 -1 0 1 2 3 4 5 6 7

**21.** Show $x \geq -5$ on the number line:

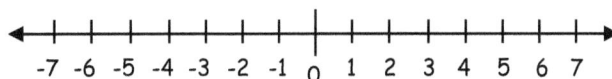

$$\longleftarrow \mid \ \mid \ \mid \ \mid \ \mid \ \mid \ \mid \ \mid \ \mid \ \mid \ \mid \ \mid \ \mid \ \mid \longrightarrow$$
-7 -6 -5 -4 -3 -2 -1 0 1 2 3 4 5 6 7

# Answer Key

## for

## MathWise Integers

## What Are Integers Anyway?

Integers are whole numbers and their opposites.

NO BETWEEN NUMBERS! (NO FRACTIONS OR NUMBERS TO THE RIGHT OF THE DECIMAL POINT)

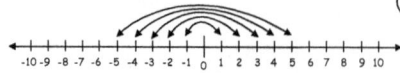

-10 -9 -8 -7 -6 -5 -4 -3 -2 -1 0 1 2 3 4 5 6 7 8 9 10

POSITIVE NUMBERS ARE USUALLY NOT WRITTEN WITH A POSITIVE SIGN (+2, ETC.)!

**Circle the integers (positive or negative numbers—no between fractions or decimals)**

1. ⑦  $\frac{1}{2}$  .3  ⑨
2. -.4  $\frac{3}{4}$  ⟨-2⟩  ⟨+5⟩
3. $-3\frac{1}{2}$  ⑫  ⟨-6⟩  $\frac{6}{5}$

**Give the opposite of each integer**

4. 7  **-7**   — SAME NUMBER, JUST PUT ON A NEGATIVE SIGN...THE OPPOSITE OF POSITIVE IS MINUS!

5. -3  **3**   — THE OPPOSITE OF A NEGATIVE IS A POSITIVE, SO JUST TAKE OFF THE SIGN!

6. -5  **5**

7. -x  **x**

8. y  **-y**

9. -a  **a**

**Put a dot on each number and its opposite on the number line; connect them with arrows**

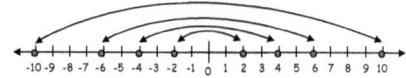

-10 -9 -8 -7 -6 -5 -4 -3 -2 -1 0 1 2 3 4 5 6 7 8 9 10

10. 2     11. -4     12. 6     13. -10

This one is done for you

© Peter Wise, 2014

1

---

## Subtraction as Adding Negatives

Any time you subtract, this is the same as adding a negative number   $10 - 3 \rightarrow 10 + (-3)$

'MINUS' IS THE SAME AS 'PLUS A NEGATIVE'!

REWRITE -3 AS + (-3)

EVERY TIME YOU SUBTRACT, IT'S THE SAME AS ADDING A NEGATIVE NUMBER!

**Rewrite the following subtraction problems as adding negative integers**

1. $5 - 2 \rightarrow$ **5 + (-2)**   (Right now just focus on rewriting, rather than solving the problem)
2. $6 - 5 \rightarrow$ **6 + (-5)**
3. $7 - 4 \rightarrow$ **7 + (-4)**
4. $12 - 6 \rightarrow$ **12 + (-6)**
5. $10 - 3 \rightarrow$ **10 + (-3)**
6. $14 - 7 \rightarrow$ **14 + (-7)**
7. $20 - 5 \rightarrow$ **20 + (-5)**
8. $15 - 10 \rightarrow$ **15 + (-10)**

9. $-16 - 6 \rightarrow$ **-16 + (-6)**
10. $-8 - 2 \rightarrow$ **-8 + (-2)**
11. $13 - 7 \rightarrow$ **13 + (-7)**
12. $19 - 11 \rightarrow$ **19 + (-11)**
13. $-30 - 5 \rightarrow$ **-30 + (-5)**
14. $a - b \rightarrow$ **a + (-b)**
15. $x - y \rightarrow$ **x + (-y)**

BACKWARDS:
Now write as a subtraction problem

16. $a + (-b) \rightarrow$ **a - b**

17. $\frac{2}{3} + \left(-\frac{1}{2}\right)$   **$\frac{2}{3} - \frac{1}{2}$**

© Peter Wise, 2014

2

---

## Same Sign Rule

SAME SIGN RULE "SSR"
• ADD THE NUMBERS
• ADD THE SIGN!

IF ALL THE NUMBERS ARE NEGATIVE, JUST ADD UP ALL OF THE NUMBERS AND PUT ON A NEGATIVE SIGN!

IF ALL THE NUMBERS HAVE THE SAME SIGN, JUST GIVE YOUR ANSWER THE SAME SIGN AND ADD UP ALL THE NUMBERS.

With same signs you always ADD the numbers

2 NEGATIVE ONES + 3 NEGATIVE ONES = HOW MANY NEGATIVE ONES?

**Add or subtract the following integers**

1. $-2 + (-3) =$ **-5**   JUST ADD UP THE NUMBERS AS THOUGH THEY ARE POSITIVE, AND MAKE YOUR ANSWER NEGATIVE AT THE END!

SINCE BOTH NUMERS ARE NEGATIVE, YOUR ANSWER WILL BE NEGATIVE!

BOTH INTEGERS ARE NEGATIVE SO THE ANSWER WILL BE NEGATIVE.

2. $-4 + (-6) = -$ **10**   4 + 6 GOES HERE (YOU KNOW THAT THE ANSWER IS NEGATIVE BECAUSE EVERYTHING YOU ARE ADDING IS NEGATIVE!

3. $-5 + (-2) = -$ **7**   IF YOU TAKE AWAY 5 (-5) AND YOU TAKE AWAY 2 (-2) HOW MANY DID YOU REALLY TAKE AWAY?

4. $6 + 3 = +$ **9**   THIS TIME BOTH NUMBERS ARE POSITIVE, SO THE ANSWER WILL BE POSITIVE!

5. $-7 + (-7) =$ **-14**

6. $-7 - 8 =$ **-15**   THIS IS REALLY THE SAME AS SUBTRACTING 15 IN TWO STAGES

7. $-5 - 1 =$ **-6**

8. $-10 + (-2) =$ **-12**

9. $-1 + (-1) + (-1) =$ **-3**

10. $-2 + (-2) + (-2) =$ **-6**

11. $-7 - 2 - 3 =$ **-12**

12. $-4 - 8 + (-2) =$ **-14**

© Peter Wise, 2014

3

## You Always Add Same Sign Integers

Example

If integers have the same sign (either positive or negative) you ALWAYS ADD them. After you have added them, just make sure you PUT ON A NEGATIVE sign if the numbers you added are NEGATIVE.

(You don't normally put a positive sign on positive numbers. We assume that if a number has no sign, it's positive.)

**Add negatives together like they're positive; then put on a negative sign**

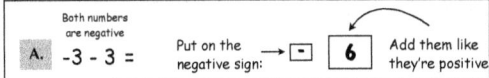

Both numbers are negative

A. $-3 - 3 =$ Put on the negative sign: $\boxed{-}$ $\boxed{6}$ Add them like they're positive

**Smaller negatives when added equal a larger negative**

ADD these SAME SIGN integers; remember to put on the right sign!

1. $-4 - 4 = \boxed{-8}$  *IF YOU TAKE AWAY 4 TWICE YOU GET THIS!*
2. $4 + 4 = \boxed{8}$  *IF YOU ADD 4 TWICE YOU GET THIS!*
3. $-4 + (-10) = \boxed{-14}$
4. $(+7) + (+8) = \boxed{15}$
5. $(-5) + (-1) = \boxed{-6}$  *TAKE AWAY ONE MORE THAN 5!*
6. $(-5) + (-5) = \boxed{-10}$  *SAME AS 2 TIMES (-5)!*

13. $6 + 7 = \boxed{13}$
14. $(-8) - 10 = \boxed{-18}$
15. $(-2) - 3 = \boxed{-5}$
16. $-2 + (-14) = \boxed{-16}$
17. $17 + 3 = \boxed{20}$
18. $-9 + (-3) = \boxed{-12}$

© Peter Wise, 2014

4

---

## Same Sign Integers

Solve these same sign integer problems

1. $-5 - 5 = \boxed{-10}$
2. $-3 + (-6) = \boxed{-9}$
3. $(-2) + (-5) = \boxed{-7}$
4. $3 + 8 = \boxed{11}$
5. $(+4) + (+6) = \boxed{10}$
6. $(-2) + (-2) = \boxed{-4}$
7. $17 + 17 = \boxed{34}$
8. $(-17) + (-17) = \boxed{-34}$
9. $14 + (+3) = \boxed{17}$
10. $-2 - 13 = \boxed{-15}$
11. $-4 + (-4) = \boxed{-8}$
12. $12 + (+13) = \boxed{25}$

13. $-10 + (-4) = \boxed{-14}$
14. $(-5) - 7 = \boxed{-12}$
15. $14 + 8 = \boxed{22}$
16. $-12 + (-2) = \boxed{-14}$
17. $16 + 5 = \boxed{21}$
18. $(-7) - 9 = \boxed{-16}$
19. $10 + 14 = \boxed{24}$
20. $-4 - 16 = \boxed{-20}$
21. $135 + 135 = \boxed{270}$
22. $-100 + (-47) = \boxed{-147}$
23. $-65 + (-20) = \boxed{-85}$
24. $-135 - 20 = \boxed{-155}$

© Peter Wise, 2014

5

---

## Combining Negative Numbers

Example

A group of SMALLER negative numbers can be combined to make one BIGGER negative number!

This is really just subtracting the same amount in two stages!

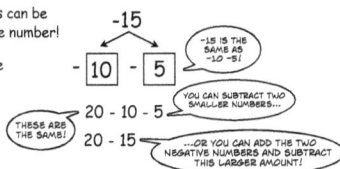

$-15$
$- \boxed{10} \quad - \boxed{5}$  *-15 IS THE SAME AS -10 -5!*

*YOU CAN SUBTRACT TWO SMALLER NUMBERS...*
$20 - 10 - 5$  *THESE ARE THE SAME!*
$20 - 15$  *...OR YOU CAN ADD THE TWO NEGATIVE NUMBERS AND SUBTRACT THIS LARGER AMOUNT!*

Break up the negative number into smaller negative numbers

1. $-5$
$8 - \boxed{3} - \boxed{2} = \boxed{3}$

2. $-7$
$10 - \boxed{2} - \boxed{5} = \boxed{3}$

3. $-10$  **answers vary**
$20 - \boxed{} - \boxed{} = \boxed{10}$  *PICK ANY TWO NUMBERS THAT ADD UP TO 10!*

4. $-20$
$25 - \boxed{} - \boxed{} = \boxed{5}$
**answers vary**

5. $-12$
$18 - \boxed{} - \boxed{} = \boxed{6}$
**answers vary**

6. $-6$
$10 - \boxed{} - \boxed{} = \boxed{4}$
**answers vary**

7. $-9$
$12 - \boxed{} - \boxed{} = \boxed{3}$
**answers vary**

8. $-14$
$16 - \boxed{} - \boxed{} = \boxed{2}$
**answers vary**

9. $-10$
$14 - \boxed{} - \boxed{} - \boxed{} = \boxed{4}$  *PICK ANY THREE NUMBERS THAT ADD UP TO 10!*

© Peter Wise, 2014

6

---

## Making One Large Negative Number

Example

When you have a group of negative numbers, add them up and make one BIG negative!

$10 \boxed{-2 \ -2 \ -2}$  *IF YOU SUBTRACT 2 THREE TIMES, IT IS THE SAME AS SUBTRACTING 6!*
*ADD THEM UP LIKE THEY ARE POSITIVE, BUT PUT ON A NEGATIVE SIGN!*
$10 - 6$

Remember that $-a$ and $+ (-a)$ are the same!

Combine the smaller negatives into one big negative number; then solve

1. $-5 - 4$
$= - \boxed{9}$  *ADD UP THE NEGATIVES LIKE THEY ARE POSITIVE...JUST PUT ON THE NEGATIVE SIGN AFTERWARDS!*
$10 - 5 - 4$  *TWO SMALLER NEGATIVES MAKE ONE LARGER NEGATIVE!*
$= 10 - \boxed{9} = \boxed{1}$

2. $-3 - 3$
$= - \boxed{6}$
$10 - 3 - 3$
$= 10 - \boxed{6} = \boxed{4}$

3. $-5 - 5 - 2$
$= - \boxed{12}$
$20 - 5 - 5 - 2$
$= 20 - \boxed{12} = \boxed{8}$

4. $10 - 2 - 2$
$10 - \boxed{4} = \boxed{6}$

5. $16 - 1 + (-1)$
$16 - \boxed{2} = \boxed{14}$

6. $7 - 1 - 2$
$7 - \boxed{3} = \boxed{4}$

7. $12 - 2 + (-3)$
$12 - \boxed{5} = \boxed{7}$

8. $13 + (-3) + (-6)$
$13 - \boxed{9} = \boxed{4}$

© Peter Wise, 2014

7

## Same Sign Integers

Solve these same sign integer problems

1. $8 + 8 = $ **16**

2. $-8 + (-8) = $ **-16**

_8_ neg steps + _8_ neg steps = _16_ neg steps

3. $(-3) + (-1) = $ **-4**

take away _3_ and take away _1_

= take away _4_

4. $2 + 3 = $ **5**

add _2_ and add _3_

= _5_

5. $(+7) + (+3) = $ **10**

6. $(-4) + (-4) = $ **-8**

7. $-2 + (-2) = $ **-4**

8. $16 + 16 = $ **32**

9. $(-16) + (-16) = $ **-32**

10. $13 + (+15) = $ **28**

13. $-5 + (-3) = $ **-8**

_5_ steps LEFT on a number line + _3_ steps LEFT on a number line = _8_ steps LEFT on a number line

14. $(-6) - 1 = $ **-7**

15. $8 + 9 = $ **17**

16. $-1 + (-5) = $ **-6**

17. $14 + 7 = $ **21**

18. $(-12) - 8 = $ **-20**

19. $13 + 7 = $ **20**

20. $-8 - 5 = $ **-13**

21. $-50 - 50 = $ **-100**

© Peter Wise, 2014

8

---

## Calculating with Groups of Negatives

Example

A. $9 - 2 - 3 = $ **4**

minus 5

pos. → **9**

neg's → **-5**

YOU CAN SUBTRACT IN TWO STEPS (-2 AND THEN -3), BUT HERE COMBINE BOTH NEGATIVES TO MAKE ONE LARGER NEGATIVE!

REMEMBER! IF IT DOESN'T HAVE A SIGN, IT'S POSITIVE!

Add/subtract the following integers

1. $10 - 3 - 1 = $ **6**

(10) pos. → **10**

(-3 - 1) neg's → **-4**

ADD THE NEGATIVES TO MAKE LARGER NEGATIVE!

Add negatives like the are positive, then put on a negative sign

2. $-3 + 20 - 2 = $ **15**

pos. → **20**

neg's → **-5** (-3 - 2)

3. $-2 - 2 - 2 + 18 = $ **12**

pos. → **18**

neg's → **-6**

4. $-4 + 16 + (-7) = $ **5**

pos. → **16**

neg's → **-11**

5. $-5 + 14 - 3 = $ **6**

pos. → **14**

neg's → **-8**

6. $-6 - 3 + 12 = $ **3**

pos. → **12**

neg's → **-9**

7. $20 - 2 - 5 = $ **13**

pos. → **20**

neg's → **-7**

8. $-8 + 15 - 2 = $ **5**

pos. → **15**

neg's → **-10**

9. $-7 + 18 - 7 = $ **4**

pos. → **18**

neg's → **-14**

10. $15 - 3 - 4 = $ **8**

pos. → **15**

neg's → **-7**

© Peter Wise, 2014

9

---

## Calculating with Groups of Negatives, pt. 2

Add/subtract the following integers

1. $30 - 5 - 10 = $ **15**

pos. → **30**

neg's → **-15**

ADD THE NEGATIVES TO MAKE ONE LARGER NEGATIVE!

2. $-2 + 28 - 4 = $ **22**

pos. → **28**

neg's → **-6**

3. $-7 + 19 + (-5) = $ **7**

pos. → **19**

neg's → **-12**

4. $-5 - 9 + 17 = $ **3**

pos. → **17**

neg's → **-14**

5. $-6 + 14 + (-7) = $ **1**

pos. → **14**

neg's → **-13**

6. $24 + (-8) + (-4) = $ **12**

pos. → **24**

neg's → **-12**

7. $15 + (-5) + (-6) = $ **4**

pos. → **15**

neg's → **-11**

8. $(-6) + 11 + (-7) = $ **-2**

pos. → **11**

neg's → **-13**

9. $-2 + (-4) + (-3) + 14 = $ **5**

pos. → **14**

neg's → **-9**

10. $-3 + 2 - 8 + 4 = $ **-5**

pos. → **6**

neg's → **-11**

11. $-2 + (4) - 6 + (10) = $ **6**

pos. → **14**

neg's → **-8**

12. $(3) - 5 + (8) - 9 = $ **-3**

pos. → **11**

neg's → **-14**

© Peter Wise, 2014

10

---

## Concept Quiz

1. Integers are __**whole**__ numbers and their __**opposites**__.

2. Whenever you SUBTRACT, it's the same as __**add**__ing a __**negative**__ number.

3. Rewrite 10 - 6 as an addition problem: __**10 + (-6)**__

4. If integers have SAME SIGNS (all positive numbers or all negative numbers) you always (ADD) SUBTRACT the numbers before putting on the same sign.
   (circle one)

5. When adding a group of smaller negative numbers you always get a negative number that is CLOSER (FARTHER) from zero.
   (circle one)

6. Rewrite -4 - 4 as an addition problem: __**- 4 + (-4)**__ Solve it: __**-8**__

7. Rewrite 10 - 2 - 6 as a problem with only two numbers: __**10 – 8**__

8. If you took away 3 apples and took away another 2 apples how many apples did you really take away?
   __**five apples**__

© Peter Wise, 2014

11

## Opposite Sign Rule

**Example**

| | THE NUMBER WITH THE NEGATIVE SIGN IS GREATER THAN THIS NUMBER THAT IS POSITIVE! | Make both numbers positive | Find the difference (subtract) | Find the sign of the higher number; give your answer this sign | Answer: |
|---|---|---|---|---|---|
| A. | -12 + 10 | 12  10 | 2 | - | -2 |

**Follow the steps and fill in the boxes to get your answers**

ADDITION RULE FOR OPPOSITE SIGN INTEGERS:

*PRETEND THEY'RE POSITIVE...  SUBTRACT...  ADD THE WINNING SIGN!*

| | | Make both numbers positive | Find the difference (subtract) | Find the sign of the higher number; give your answer this sign | Answer: |
|---|---|---|---|---|---|
| 1. | 6 + (-7) | 6  7 | 1 | - | -1 |

*AFTER YOU MAKE BOTH NUMBERS POSITIVE, SUBTRACT THE SMALLER NUMBER FROM THE LARGER NUMBER!*

| | | Make both numbers positive | Find the difference (subtract) | Find the sign of the higher number; give your answer this sign | Answer: |
|---|---|---|---|---|---|
| 2. | -8 + 5 | 8  5 | 3 | - | -3 |
| 3. | -10 + 4 | 10  4 | 6 | - | -6 |
| 4. | 7 - 12 | 7  12 | 5 | - | -5 |
| 5. | -6 + 17 | 6  17 | 11 | + | 11 |

© Peter Wise, 2014

12

---

## Opposite Sign Rule

**Follow the steps and fill in the boxes to get your answers**

| | | Make both numbers positive | Find the difference (subtract) | Find the sign of the higher number; give your answer this sign | Answer: |
|---|---|---|---|---|---|
| 1. | -7 + 3 | 7  3 | 4 | - | -4 |
| 2. | 4 + (-9) | 4  9 | 5 | - | -5 |
| 3. | -14 + 10 | 14  10 | 4 | - | -4 |
| 4. | (-3) + 9 | 3  9 | 6 | + | 6 |
| 5. | -18 + 20 | 18  20 | 2 | + | 2 |
| 6. | 6 + (-9) | 6  9 | 3 | - | -3 |

ADDITION RULE FOR OPPOSITE SIGN INTEGERS:

*PRETEND THEY'RE POSITIVE...  SUBTRACT...  ADD THE WINNING SIGN!*

© Peter Wise, 2014

13

---

## Opposite Sign Rule

**Follow the steps and fill in the boxes to get your answers**

| | | Make both numbers positive | Find the difference (subtract) | Find the sign of the higher number; give your answer this sign | Answer: |
|---|---|---|---|---|---|
| 1. | 6 + (-8) | 6  8 | 2 | - | -2 |
| 2. | -6 + 8 | 6  8 | 2 | + | 2 |
| 3. | -5 + 2 | 5  2 | 3 | - | -3 |
| 4. | (-9) + 10 | 9  10 | 1 | + | 1 |
| 5. | 12 + (-20) | 12  20 | 8 | - | -8 |
| 6. | -30 + 24 | 30  24 | 6 | - | -6 |
| 7. | 16 - 19 | 16  19 | 3 | - | -3 |

© Peter Wise, 2014

14

---

## Adding Integers on a Number Line

**Example**

Negatives go LEFT     Positives go RIGHT

A. 2 + -5

start at 0
go RIGHT 2   go LEFT 5

**Add the integers by drawing arrows on the number line; put a dot on the answer**

Negatives go LEFT     Positives go RIGHT

*IMPORTANT! Same signs go in the same direction   That's why you ADD them!*

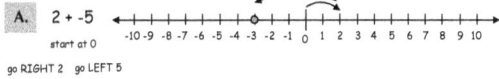

1. 2 + (-3)
start at 0
go RIGHT 2   go LEFT 3

**Notice!** If the negative number is bigger than the positive number you will be left of zero! (negative)

2. -4 + (7)
start at 0

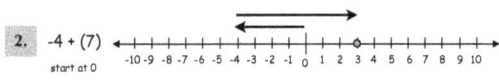

3. -1 + (5)
start at 0

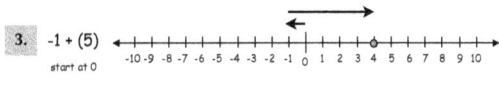

4. -3 -6
start at 0

SAME AS + (-6)

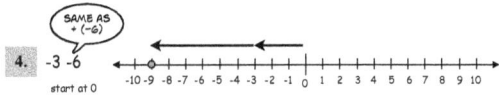

5. -8 + 7
start at 0

*Opposite signs go in the opposite direction   That's why you SUBTRACT them!*

© Peter Wise, 2014

15

## Adding Integers on a Number Line

**Example**

**A.** -3 - 6

Negatives go LEFT        Positives go RIGHT

start at 0

go LEFT 3    go LEFT 6 more steps

Add the integers by drawing arrows on the number line; put a dot on the answer

Negatives go LEFT        Positives go RIGHT

**1.** -2 + 7

start at 0

go LEFT 2    go RIGHT 7

**2.** -5 + (-2)

start at 0

go LEFT 5    go LEFT two more

**Notice!** If all integers are negative, you just keep going left

**3.** -3 + (-3)

start at 0

**Notice!** Your answer is the same distance away from zero as it would be if the answer were POSITIVE

**4.** 8 + (-2)

start at 0

**5.** -4 + (-5)

start at 0

16

---

## Visualizing Integers as Counters

$\bigcirc N$ = negative 1      $\bigcirc P$ = positive 1

**OPPOSITE SIGN INTEGERS**

**A.** (-3) + 4 = 1

ONE POSITVE IS LEFT, SO THE ANSWER IS +1!

Cancel out one N for every P
See what is left!

**SAME SIGN INTEGERS**

**B.** (-3) + (-2) = -5

3 negatives + 2 negatives = five negatives (-5)!

NOTHING TO CANCEL OUT BECAUSE EVERYTHING IS THE SAME SIGN!

Add the following integers using COUNTERS

**1.** (-2) + 1 = **-1**

CANCEL OUT A NEGATIVE AND POSITIVE!

**2.** (-1) + 2 = **1**

LABEL THE CIRCLES; THEN CANCEL OUT A NEGATIVE AND A POSITIVE!

**3.** -1 + (-1) = **-2**

**4.** 2 + (-3) = **-1**

**5.** -2 + 3 = **1**

**6.** (-3) + 3 = **0**

**7.** (-4) + 3 = **-1**

**8.** (-4) + (-3) = **-7**

**9.** (-2) + (-2) + 1 = **-3**

**10.** -1 + 3 + (-4) = **-2**

17

---

## Enrichment: Substitution with Integers

Use substitution to solve the following problems

**Example**

**A.** a + b = [ ]    (-2) + (-5) = **-7**

a = -2    b = -5

**1.** a - b = ( **7** ) - ( **3** ) = **4**

a = 7    b = 3

**2.** -a + b = ( **-5** ) + ( **-6** ) = **-11**

a = 5    b = -6

**3.** -a + b = ( **-10** ) + ( **-3** ) = **-13**

a = 10    b = -3

**4.** -x + (-y) = ( **-3** ) + ( **-2** ) = **-5**

x = 3    y = 2

**5.** x + (-y) = ( **6** ) + ( **-9** ) = **-3**

x = 6    y = 9

*A LITTLE MORE CHALLENGING...*

**6.** a + b + c = ( **-2** ) + ( **10** ) + ( **-6** ) = **2**

a = -2    b = 10    c = -6

**7.** x + y - z = **-4 + 20 - 5** = **11**

x = -4    y = 20    z = 5

*REPEATED ADDITION OF INTEGERS!*

**8.** b + b = **-8**

b = -4

**9.** a + a + a = **-9**

a = -3

**10.** y + y + y = **-18**

y = -6

**11.** x + x + x + x = **-20**

x = -5

18

---

## Elevator-Style Number Lines

**Example**

**A.** 2 + (-3)

POSITIVE GOES UP!

NEGATIVE GOES DOWN!

START AT ZERO!

up 2

down -3

1. Start at 0
2. Go UP 2
3. Go DOWN 3

Draw up and down arrows for the integers; draw a dot on the final answer

Positives go UP        Negatives go DOWN

**1.** 3 + (-4)

**2.** -7 + 8

**3.** -3 + (-3)

**4.** 6 + (-3)

**5.** -2 + (-5)

NOTICE! THE ARROWS GO IN OPPOSITE DIRECTIONS WHENEVER YOU HAVE ONE POSITIVE AND ONE NEGATIVE INTEGER!

THIS IS WHY YOU SUBTRACT THESE INTEGERS!

19

## Elevator-Style Number Lines

Draw up and down arrows for the integers; draw a dot on the final answer

**Positives go UP**   **Negatives go DOWN**

1. 5 + (-6)
2. 6 + (-5)
3. -3 + (-5)
4. -3 + (-4)
5. 8 + (-2)

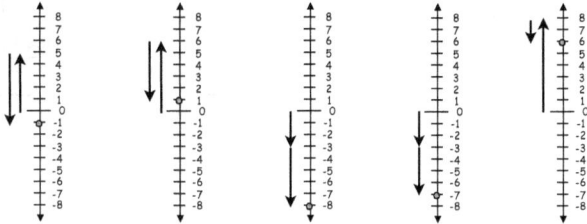

6. -4 + 5
7. 5 + -4
8. -2 + 10
9. -2 + (-3)
10. 5 + (-7)

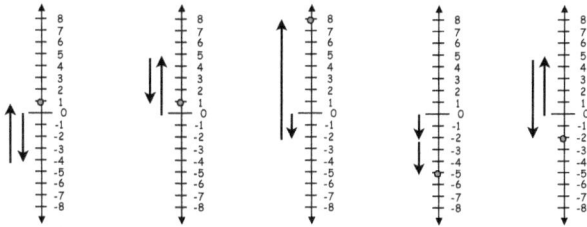

20

---

## Adding Groups of Integers

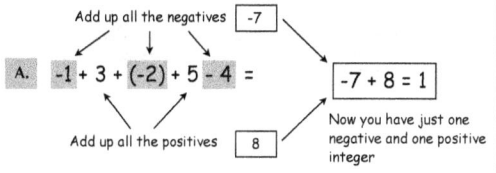

Example

A. -1 + 3 + (-2) + 5 - 4 =

Add up all the negatives → -7

Add up all the positives → 8

-7 + 8 = 1

Now you have just one negative and one positive integer

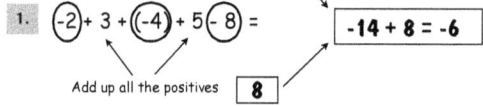

1. (-2) + 3 + ((-4)) + 5 - (8) =

Circle and add up all the negatives → -14
Add them like they are positive, and then put on a negative sign

Add up all the positives → 8

-14 + 8 = -6

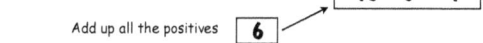

2. -5 + 3 - 5 + 3 =

Circle and add up all the negatives → -10

Add up all the positives → 6

-10 + 6 = -4

3. 6 + (-7) + 6 + (-2) =

Circle and add up all the negatives → -9

Add up all the positives → 12

12 - 9 = 3

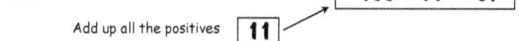

4. 5 + (-50) + 6 + (-50) =

Circle and add up all the negatives → -100

Add up all the positives → 11

-100 + 11 = -89

21

---

## Adding Groups of Integers

1. -2 + 5 - 2 + 5 = **6**

Negatives = **-4** Positives = **10**
combine all the negatives · combine all the positives · now solve

2. -4 + 10 + (-4) + 10 = **12**

Negatives = **-8** Positives = **20**
combine all the negatives · combine all the positives

3. -3 + 20 + (-3) + 20 = **34**

Negatives = **-6** Positives = **40**

4. 4 + (-10) + 4 + (-10) = **-12**

Negatives = **-20** Positives = **8**

5. 7 - 4 + 6 - 5 = **4**

Negatives = **-9** Positives = **13**

6. -12 + 4 + (-7) + 8 = **-7**

Negatives = **-19** Positives = **12**

7. 4 + (-6) + 14 - 2 = **10**

Negatives = **-8** Positives = **18**

8. -12 + 2 - 3 + 7 = **-6**

Negatives = **-15** Positives = **9**

9. 10 + (-7) + 11 + (-12) = **2**

Negatives = **-19** Positives = **21**

10. -4 + (-5) + (-6) + 7 + 8 = **0**

Negatives = **-15** Positives = **15**

11. -6 + 8 + 7 + 3 + (-6) = **6**

Negatives = **-12** Positives = **18**

12. -12 + 7 - 10 + 6 - 3 + 2 = **-10**

Negatives = **-25** Positives = **15**

22

---

## Adding and Subtracting Coefficients

**Coefficient = number touching (multiplying) a letter**

Examples

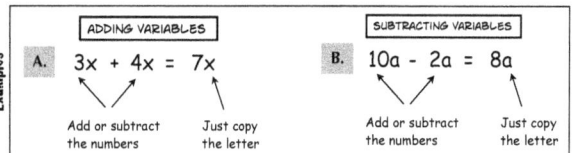

| ADDING VARIABLES | SUBTRACTING VARIABLES |
|---|---|
| A. 3x + 4x = 7x | B. 10a - 2a = 8a |

Add or subtract the numbers · Just copy the letter   Add or subtract the numbers · Just copy the letter

1. 2x + 2x = **4x**
2. 8x - 3x = **5x**
3. 3a + 10a = **13a**
4. 12n - 3n = **9n**
5. 2y + 2y + 2y = **6y**
6. 7x + 10x = **17x**
7. 5r + 8r = **13r**
8. 14x - 8x = **6x**
9. 15n - 6n = **9n**
10. 13m - 7m = **6m**

11. 3a - 4a = **-1a** or **-a**
12. 10y + (-12y) = **-2y**
13. -3n + 5n = **2n**
14. -2y + (-2y) = **-4y**
15. 6s + (-7s) = **-1s**
16. -6x + (-6x) = **-12x**
17. -8x + (-3x) = **-11x**
18. 16a - 14a = **2a**
19. -20y + 25y = **5y**
20. 17n - 19n = **-2n**

23

## Concept Quiz

1. If integers have OPPOSITE SIGNS (one positive and one negative) you always ADD / (SUBTRACT) the numbers before putting on the "winning" sign.
   (circle one)

2. According to the OPPOSITE SIGN RULE, you first pretend that both the positive and negative numbers are:

   __**positive**__

3. If the NEGATIVE number is larger than the POSITIVE number, you know that the answer will be POSITIVE / (NEGATIVE)
   (circle one)

4. On a number line negative numbers go to the (LEFT) / RIGHT
   (circle one)

5. Add / subtract the following integers by drawing arrows on the number line; put a dot on the answer:

   $-2 + (5)$
   
   -10 -9 -8 -7 -6 -5 -4 -3 -2 -1 0 1 2 3 4 5 6 7 8 9 10

6. On a number line, integers with the SAME SIGN always go in (THE SAME) / OPPOSITE directions
   (circle one)

7. Circle and add up all the negatives → **-12**

   $-6 + 3 - 6 + 7 =$ → **-2**

   Add up all the positives **10**   (put answer here)

© Peter Wise, 2014

24

---

## Double Negatives = Positive (or +)

Double negative = positive

$3 - (-4) = +7$
THE TWO NEGATIVES TOUCHING BECOME A PLUS!

Warning!
$-3 + (-4) = -7$
THESE NEGATIVES ARE NOT TOUCHING. THEY DO NOT TURN TO A PLUS!

– touching – equals +

negative touching negative = positive     negative plus a negative = negative

**Solve these integer problems; some have double negatives**

EVERY TWO NEGATIVES TOUCHING EACH OTHER CANCEL BACH OTHER OUT!

1. $2 - (-4) =$ **6**
2. $5 - (-6) =$ **11**
3. $10 - (-3) =$ **13**
4. $10 - 3 =$ **7**
5. $-(-12) + 3 =$ **15**
   SAME AS POSITIVE!
6. $16 - (-4) =$ **20**
7. $9 - (-5) =$ **14**

8. $-1 - (-2) =$ **1**
9. $-3 - (-3) =$ **0**
10. $5 - (-7) =$ **12**
11. $-2 - (-2) =$ **0**
12. $-2 - 2 =$ **-4**
13. $-(-3) - (-3) =$ **6**
14. $(-10) - (-10) =$ **0**
15. $-(-10) - (-10) =$ **20**

© Peter Wise, 2014

25

---

## Double Negatives = Positive (or +)

Example

Negative touching negative turns to PLUS

A. $3 - (-2) = 5$     touching = +
negative sign   negative sign   one negative sign rotated and joined to the other!

THINK OF A PLUS SIGN AS TWO NEGATIVE SIGNS PUT TOGETHER!

1. $10 - (-2) =$ **12**
2. $10 - 2 =$ **8**
3. $7 - (-5) =$ **12**
4. $7 - 5 =$ **2**
5. $12 - (-4) =$ **16**
6. $14 - (-6) =$ **20**
7. $14 - 6 =$ **8**
8. $-14 - 6 =$ **-20**
9. $20 - (-7) =$ **27**
10. $20 - 7 =$ **13**

11. $-(-8) - 4 =$ **4**
12. $-8 - 4 =$ **-12**
13. $17 - (-4) =$ **21**
14. $-(-11) - 3 =$ **8**
15. $15 - (-6) =$ **21**
16. $2 - (-7) =$ **9**
17. $-(-5) - (-3) =$ **8**
18. $-6 - (-7) =$ **1**
19. $-5 - (-5) =$ **0**
20. $-12 - (-10) =$ **-2**

© Peter Wise, 2014

26

---

## Double Negatives = Positive (or +)

Example

Negative touching negative turns to PLUS

A. $-2 - (-3) = 1$     touching = +
negative sign   negative sign   one negative sign rotated and joined to the other!

SAME AS -2 + 3

THINK OF A PLUS SIGN AS TWO NEGATIVE SIGNS PUT TOGETHER!

1. $8 - (-3) =$ **11**
2. $8 - 3 =$ **5**
3. $15 - (-5) =$ **20**
4. $15 - 5 =$ **10**
5. $9 - (-3) =$ **12**
6. $18 - (-4) =$ **22**
7. $19 - 6 =$ **13**
8. $14 - (-6) =$ **20**
9. $22 - (-3) =$ **25**
10. $22 - 3 =$ **19**

11. $-3 - (-5) =$ **2**
12. $-4 - (-12) =$ **8**
13. $19 - (-5) =$ **24**
14. $8 - (-8) =$ **16**
15. $-(-15) - 15 =$ **0**
16. $-2 - (-9) =$ **7**
17. $-3 - (-11) =$ **8**
18. $-(-8) - (-6) =$ **14**
19. $-5 - (-6) =$ **1**
20. $-(-14) - 12 =$ **2**

© Peter Wise, 2014

27

## 28

## Review: Opposite Sign Integers

**Follow the steps; add or subtract the following integers**

p = positive    n = negative

A. SIGN and the
B. NUMBER

1. Determine the SIGN your answer will be    2. Determine the NUMBER

**OPPOSITE sign integers always SUBTRACT!**

**Example**

A.  ⓟn  ⓟn   answer will be
    6 + (-7)   p ⓝ

| 7, 6 | ← Make both numbers positive. |
| 7 - 6 | ← Subtract the smaller number from the larger number. |

Answer:  − 1

1.  ⓟn  pⓝ   answer will be
    10 - 13   p ⓝ

| 10, 13 | ← Make both numbers positive. |
| 13-10 | ← Subtract the smaller number from the larger number. |

Answer:  − 3

2.  ⓟn  ⓟn   answer will be
    -7 + (12)   ⓟ n

| 7, 12 | ← Make both numbers positive. |
| 12-7 | ← Subtract the smaller number from the larger number. |

Answer:  + 5

3.  ⓟn  pⓝ   answer will be
    37 + (-43)   p ⓝ

| 37, 43 | ← Make both numbers positive. |
| 43-37 | ← Subtract the smaller number from the larger number. |

Answer:  − 6

4.  pⓝ  ⓟn   answer will be
    -53 + (49)   p ⓝ

| 53, 49 | ← Make both numbers positive. |
| 53-49 | ← Subtract the smaller number from the larger number. |

Answer:  − 4

ADDITION RULE FOR NUMBERS WITH OPPOSITE SIGNS:

*PRETEND THEY'RE POSITIVE... SUBTRACT... ADD THE WINNING SIGN!*

© Peter Wise, 2014

---

## 29

## Integer Review

**Follow the steps; add or subtract the following integers**

1. -1 + (5)    start at 0

2. -4 - 2    start at 0

3. (-3) + 1 = **-2**

ⓝ ⓝ ⓟ   LABEL THE CIRCLES WITH EITHER N OR P!
ⓝ
THEN CANCEL OUT ONE NEGATIVE WITH ONE POSITIVE AND SEE WHAT'S LEFT!

4. (-2) + (-4) = **-6**

ⓝ ⓝ ⓝ ⓝ   MAKE CIRCLES AND PUT "P" FOR POSITIVE AND "N" FOR NEGATIVE; THEN CANCEL TO SEE WHAT'S LEFT!
ⓝ ⓝ

THE LARGER POSITIVE NUMBER MINUS THE SMALLER POSITIVE NUMBER!

**Opposite Sign Rule**

| | Make both numbers positive | Find the difference (subtract) | Find the sign of the higher number; give your answer this sign | Answer: |
|---|---|---|---|---|
| 5. -7 + 4 = | 7 \| 4 | 3 | − | -3 |
| 6. -2 + 10 = | 2 \| 10 | 8 | + | 8 |

7. -12 + 8 = **-4**    8. -3 + 5 = **2**    9. -4 + 5 = **1**

**Same Sign Rule**

10. -2 + (-2) + (-2) = **-6**    12. -4 + -5 = **-9**

11. -3 + (-10) + (-10) = **-23**    13. -2 + -10 = **-12**

© Peter Wise, 2014

---

## 30

## Integer Addition and Subtraction

**Circle the correct rule and operation, then solve**

REMEMBER THAT DOUBLE NEGATIVE = POSITIVE!

**SAME sign numbers - you always ADD** — Just give your answer the same sign as the number you added!

**OPPOSITE sign numbers - you always SUBTRACT** — Here give your answer the sign of the higher number!

ALWAYS SUBTRACT THE SMALLER NUMBER FROM THE LARGER NUMBER!

| | SSR = Same Sign Rule | OSR = Opposite Sign Rule | Will you + or − the numbers? | ANSWER: |
|---|---|---|---|---|
| 1. -3 + (-5) = | (SSR) | OSR | (add) subtract | -8 |
| 2. -4 - 2 = | (SSR) | OSR | add (subtract) | -6 |
| 3. 10 + (-6) = | SSR | (OSR) | add (subtract) | 4 |
| 4. 8 - (-4) = | (SSR) | OSR | (add) subtract | 12 |
| 5. -6 - (-2) = | SSR | (OSR) | add (subtract) | -4 |
| 6. -18 + 18 = | SSR | (OSR) | add (subtract) | 0 |
| 7. -10 - (-3) = | SSR | (OSR) | add (subtract) | -7 |
| 8. -(-4) + 5 = | (SSR) | OSR | (add) subtract | 9 |
| 9. -3 + 4 = | SSR | (OSR) | add (subtract) | 1 |
| 10. -2 - 7 = | (SSR) | OSR | (add) subtract | -9 |
| 11. 15 - 22 = | SSR | (OSR) | add (subtract) | -7 |
| 12. -4 + (-16) = | (SSR) | OSR | (add) subtract | -20 |

© Peter Wise, 2014

---

## 31

## Integer Addition and Subtraction

**Circle the correct rule and operation, then solve**

| | SSR = Same Sign Rule | OSR = Opposite Sign Rule | Will you + or − the numbers? | ANSWER: |
|---|---|---|---|---|
| 1. -5 + 4 = | SSR | (OSR) | add (subtract) | -1 |
| 2. 4 - 5 = | SSR | (OSR) | add (subtract) | -1 |
| 3. -6 + (-3) = | (SSR) | OSR | (add) subtract | -9 |
| 4. -7 + 7 = | SSR | (OSR) | add (subtract) | 0 |
| 5. -20 + (-1) = | (SSR) | OSR | (add) subtract | -21 |
| 6. -10 + 2 = | SSR | (OSR) | add (subtract) | -8 |
| 7. -10 - 10 = | (SSR) | OSR | (add) subtract | -20 |
| 8. -4 + (-5) = | (SSR) | OSR | (add) subtract | -9 |
| 9. 10 + (-11) = | SSR | (OSR) | add (subtract) | -1 |
| 10. 8 - (-2) = | (SSR) | OSR | (add) subtract | 10 |
| 11. -3 - (-10) = | SSR | (OSR) | (add) subtract | 7 |
| 12. -(-5) + 3 = | (SSR) | OSR | (add) subtract | 8 |

© Peter Wise, 2014

## Integer Addition and Subtraction

### Add or subtract the following integers

1. 3 - 4 = **-1**
2. -4 + 3 = **-1**
3. -3 - 4 = **-7**
4. -10 - 6 = **-16**
5. -2 + (-3) = **-5**
6. 7 - (-5) = **12**
7. -9 + (-10) = **-19**

8. -6 + (-3) = **-9**
9. 4 + (-4) = **0**
10. -7 + 8 = **1**
11. 8 - 9 = **-1**
12. -4 + 10 = **6**
13. -17 + 15 = **-2**
14. -21 + 20 = **-1**

15. Write 15 - 4 as an addition problem with a negative integer   **15 + (-4)**

32

---

## Integer Addition and Subtraction

### Add or subtract the following integers

1. 5 - 5 = **0**
2. 5 - 6 = **-1**
3. 7 - 5 = **2**
4. 5 - 7 = **-2**
5. 10 - 2 = **8**
6. 2 - 10 = **-8**
7. 6 - 1 = **5**
8. 1 - 6 = **-5**
9. -2 - 4 = **-6**
10. 10 - 13 = **-3**
11. -6 + (-6) = **-12**
12. -10 + (-6) = **-16**

13. -10 + 6 = **-4**
14. (-4) - 8 = **-12**
15. -5 + 4 = **-1**
16. 12 - 14 = **-2**
17. -7 + (-3) = **-10**
18. -8 + (-8) = **-16**
19. 10 - 11 = **-1**
20. -5 + (-4) = **-9**
21. 20 + (-22) = **-2**
22. -13 + 3 = **-10**
23. -13 + (-3) = **-16**

WHY IS THIS? (COMPARE THIS PROBLEM TO THE FIRST ONE!)

COMPARE THESE TWO PROBLEMS!

WHY IS ONE POSITIVE AND THIS OTHER NEGATIVE?

REMEMBER! IF THE LARGER NUMBER IS POSITIVE, THE ANSWER WILL BE POSITIVE!

THESE ARE BOTH NEGATIVE NUMBERS!

HOW MANY NEGATIVES DO YOU HAVE?

SAME AS 2 - (-8)!

WHAT HAPPENS WHEN YOU SUBTRACT ONE MORE THAN YOU HAVE?

33

---

## Integer Addition and Subtraction

### Add or subtract the following integers

1. 4 + (-6) = **-2**
2. 17 + (-5) = **12**
3. -15 - 4 = **-19**
4. 6 - (-4) = **10**
5. (-4) + (-3) = **-7**
6. (-8) - (-4) = **-4**
7. 18 - 20 = **-2**
8. 2 - (-6) = **8**
9. 10 + (-13) = **-3**

10. (-20) + (-5) = **-25**
11. -8 - 6 = **-14**
12. -8 + (-6) = **-14**
13. 4 - (-7) = **11**
14. -3 - (-5) = **2**
15. 12 - (-10) = **22**
16. -4 + (-15) = **-19**
17. -10 + 12 = **2**
18. -10 + (-15) = **-25**

34

---

## Find the Missing Integers

1. -2 + **-2** = -4
2. -5 - **5** = -10
3. 3 + **(-3)** = 0
4. 3 + **(-4)** = -1
5. 3 + **(-5)** = -2
6. 4 + **(-3)** = 1
7. 10 + **(-12)** = -2
8. **9** + (-4) = 5
9. **10** + (-11) = -1

10. -4 + **4** = 0
11. -4 + **10** = 6
12. 8 + **(-8)** = 0
13. 8 + **(-10)** = -2
14. **(-6)** + (-3) = -9
15. 6 - **(-4)** = 10
16. **1** - 3 = -2
17. 5 + **(-10)** = -5
18. **6** - (-3) = 9

35

# Concept Quiz

1. 5 + (-2) can also be written as what simpler subtraction problem?   **5 - 2**

2. Integers are whole numbers and their **opposites** .

   Label the following integers with p for positive and n for negative:

3. | p | n |
   10 - 6

4. | n | n |
   -7 + (-3)

5. | n | p |
   -12 - (-4)

6. When you you add integers?   **when the signs are the same**

7. When do you **subtract** integers?   **when the signs are opposite**

8. What is the "Opposite-Sign Rule?"

   Step #1  **make both numbers positive**

   Step #2  **subtract (larger - smaller positive number)**

   Step #3  **give your answer the sign of the larger number**

9. What is the "Same-Sign Rule?"

   **Add the numbers as though they were positive**

   **Put on the sign of the numbers you added**

---

# Multiplying Integers

ODD number of negative signs in the problem: NEGATIVE ANSWER
EVEN number of negative signs in the problem: POSITIVE ANSWER

BECAUSE EVERY TWO NEGATIVES CANCEL EACH OTHER OUT!

A. $\boxed{-2 \cdot 3}$ = -6

ODD number of negative signs in the problem: $\boxed{\text{NEGATIVE ANSWER}}$

B. $-3 \cdot -4$ = 12

2 neg's cancel out

EVEN number of negative signs in the problem: $\boxed{\text{POSITIVE ANSWER}}$

**Examples**

---

**Circle EVEN or ODD and POS / NEG; give your product the correct sign**

Cancel out every two negatives, if none are left, the answer is positive

1. -5 · 3 = $\boxed{-15}$
   EVEN (ODD)   POS (NEG) answer
   number of negative signs in the problem (circle one)

5. 8 · (-3) = $\boxed{-24}$
   EVEN (ODD)   POS (NEG) answer
   number of negative signs in the problem

2. -5 · -4 = $\boxed{20}$
   (EVEN) ODD   (POS) NEG answer
   number of negative signs in the problem

   WHEN NUMBERS TOUCH, THEY TIMES!

6. (-6)(-6) = $\boxed{36}$
   (EVEN) ODD   (POS) NEG answer
   number of negative signs in the problem

3. (-1)(-1)(-1) = $\boxed{-1}$
   EVEN (ODD)   POS (NEG) answer
   number of negative signs in the problem

7. (-1)(-2)(-6) = $\boxed{-12}$
   EVEN (ODD)   POS (NEG) answer
   number of negative signs in the problem

4. (-1)(-1)(-1)(-1) = $\boxed{1}$
   (EVEN) ODD   (POS) NEG answer
   number of negative signs in the problem

8. (-2)(-2)(-3)(-2) = $\boxed{24}$
   (EVEN) ODD   (POS) NEG answer
   number of negative signs in the problem

---

# Multiplying Integers

**Circle EVEN or ODD and POS / NEG; give your product the correct sign**

Cancel out every two negatives, if none are left, the answer is positive

1. -5 · -5 = $\boxed{25}$
   (EVEN) ODD   (POS) NEG answer
   number of negative signs in the problem (circle one)

6. (-8)(-8) = $\boxed{64}$
   (EVEN) ODD   (POS) NEG answer
   number of negative signs in the problem

2. -6 · 3 = $\boxed{-18}$
   EVEN (ODD)   POS (NEG) answer
   number of negative signs in the problem

7. (-3)(-3)(3) = $\boxed{27}$
   (EVEN) ODD   (POS) NEG answer
   number of negative signs in the problem

3. (-2)(-2)(-2) = $\boxed{-8}$
   EVEN (ODD)   POS (NEG) answer
   number of negative signs in the problem

8. 4 · (-7) = $\boxed{-28}$
   EVEN (ODD)   POS (NEG) answer
   number of negative signs in the problem

4. (-2)(-2)(-2)(-2) = $\boxed{16}$
   (EVEN) ODD   (POS) NEG answer
   number of negative signs in the problem

9. (-3)(-8)(-1) = $\boxed{-24}$
   EVEN (ODD)   POS (NEG) answer
   number of negative signs in the problem

5. (-1)(-2)(-2)(-2) = $\boxed{8}$
   (EVEN) ODD   (POS) NEG answer
   number of negative signs in the problem

10. (-4)(-3)(3) = $\boxed{36}$
    (EVEN) ODD   (POS) NEG answer
    number of negative signs in the problem

---

# Multiplying Integers

**Circle EVEN or ODD and POS / NEG; give your product the correct sign**

Cancel out every two negatives, if none are left, the answer is positive

1. -4 · 8 = $\boxed{-32}$
   EVEN (ODD)   POS (NEG) answer
   number of negative signs in the problem

6. (-7)(4)(-1) = $\boxed{28}$

2. -6 · -8 = $\boxed{48}$
   (EVEN) ODD   (POS) NEG answer
   number of negative signs in the problem

7. (-2)(-3)(-5) = $\boxed{-30}$

3. 7 · (-2) = $\boxed{-14}$
   EVEN (ODD)   POS (NEG) answer
   number of negative signs in the problem

8. -12 · -2 = $\boxed{24}$

9. $(-7)^2$ = $\boxed{49}$

4. (-1)(-3)(-7) = $\boxed{-21}$
   EVEN (ODD)   POS (NEG) answer
   number of negative signs in the problem

10. (-3)(-3)(-3) = $\boxed{-27}$

multiply the numerators

11. $\dfrac{1}{-2} \cdot \dfrac{1}{3}$ = $\dfrac{\boxed{1}}{\boxed{-6}}$

multiply the denominators

5. -11 · -5 = $\boxed{55}$
   (EVEN) ODD   (POS) NEG answer
   number of negative signs in the problem

12. (-4)(-2)(7) = $\boxed{56}$

## Multiplication as Repeated Addition of Integers

**Multiplication is repeated addition**

Example

**A.**   $3 \cdot (-2) =$

$\boxed{-2} + \boxed{-2} + \boxed{-2} = \boxed{-6}$

### Show the multiplication problems as repeated addition; then solve

**1.**   $2 \cdot (-3) =$

$\boxed{-3} + \boxed{-3} = \boxed{-6}$

**5.**   $3 \cdot (-4) =$

$\boxed{-4} + \boxed{-4} + \boxed{-4} = \boxed{-12}$

**2.**   $4 \cdot (-2) =$

$\boxed{-2} + \boxed{-2} + \boxed{-2} + \boxed{-2} = \boxed{-8}$

**6.**   $2 \cdot (-8) =$

$\boxed{-8} + \boxed{-8} = \boxed{-16}$

**3.**   $2 \cdot (-5) =$

$\boxed{-5} + \boxed{-5} = \boxed{-10}$

**7.**   $4 \cdot (-5) =$

$\boxed{-5} + \boxed{-5} + \boxed{-5} + \boxed{-5} = \boxed{-20}$

Show repeated addition of (-5)

**4.**   $2 \cdot (-9) =$

$\boxed{-9} + \boxed{-9} = \boxed{-18}$

**8.**   $3 \cdot (-7) =$

$\boxed{-7} + \boxed{-7} + \boxed{-7} = \boxed{-21}$

Show repeated addition of (-7)

© Peter Wise, 2014

40

---

## Division as Backwards Multiplication

Example

**A.**   $12 \div (-6) = \boxed{-2}$

$= (-6) \cdot \boxed{-2}$

THE ANSWER TO THE BOTTOM PROBLEM TELLS YOU THE ANSWER TO THE TOP PROBLEM!

**WORK BACKWARDS!**

### Solve the following division problems by working BACKWARDS

**1.**   $18 \div (-3) = \boxed{-6}$

$= (-3) \cdot \boxed{-6}$   same number

START WITH THE BOTTOM PROBLEM!

**7.**   $-16 \div 2 = \boxed{-8}$

$= 2 \cdot \boxed{-8}$

**2.**   $-10 \div (-2) = \boxed{5}$

$= (-2) \cdot \boxed{5}$

**8.**   $-8 \div (-4) = \boxed{2}$

$= (-4) \cdot \boxed{2}$

**3.**   $-12 \div 3 = \boxed{-4}$

$= 3 \cdot \boxed{-4}$

**9.**   $20 \div (-5) = \boxed{-4}$

$= (-5) \cdot \boxed{-4}$

**4.**   $15 \div (-5) = \boxed{-3}$

$= (-5) \cdot \boxed{-3}$

**10.**   $-22 \div 2 = \boxed{-11}$

$= 2 \cdot \boxed{-11}$

**5.**   $-20 \div 2 = \boxed{-10}$

$= 2 \cdot \boxed{-10}$

**11.**   $18 \div (-9) = \boxed{-2}$

$= (-9) \cdot \boxed{-2}$

**6.**   $-9 \div (-3) = \boxed{3}$

$= (-3) \cdot \boxed{3}$

**12.**   $-24 \div (-6) = \boxed{4}$

$= (-6) \cdot \boxed{4}$

© Peter Wise, 2014

41

---

## Integer Division

Examples

EVEN number of negative signs in the problem:   **POSITIVE ANSWER**

ODD number of negative signs in the problem:   **NEGATIVE ANSWER**

2 neg's cancel out

BECAUSE EVERY TWO NEGATIVES CANCEL EACH OTHER OUT!

**A.**   $\boxed{-10 \div 2} = -5$

ODD number of negative signs in the problem:   $\boxed{\text{NEGATIVE ANSWER}}$

**B.**   $-6 \div -2 = 3$

EVEN number of negative signs in the problem:   $\boxed{\text{POSITIVE ANSWER}}$   **-6**

### Count the number of negative signs and solve the following division problems

**1.**   $-8 \div (-2) = \boxed{4}$

(EVEN)/ ODD   (POS) NEG answer
number of negative signs in the problem

**7.**   $-33 \div 3 = \boxed{-11}$

**2.**   $-9 \div (3) = \boxed{-3}$

EVEN (ODD)   POS (NEG) answer
number of negative signs in the problem

**8.**   $-5 \div (-1) = \boxed{5}$

**3.**   $-14 \div 7 = \boxed{-2}$

EVEN (ODD)   POS (NEG) answer
number of negative signs in the problem

**9.**   $42 \div (-6) = \boxed{-7}$

**4.**   $-32 \div (-8) = \boxed{4}$

(EVEN)/ ODD   (POS) NEG answer
number of negative signs in the problem

**10.**   $-36 \div -9 = \boxed{4}$

**5.**   $-27 \div -3 = \boxed{9}$

(EVEN)/ ODD   (POS) NEG answer
number of negative signs in the problem

**11.**   $-24 \div (-2) = \boxed{12}$

**12.**   $48 \div (-6) = \boxed{-8}$

© Peter Wise, 2014

42

---

## Substitution with Integers

### Use substitution to solve the following problems

**1.**   $a + b$

$(\mathbf{-2}) + (\mathbf{-5}) = \boxed{-7}$

$a = -2 \quad b = -5$

**3.**   $-a + b$   **-1**

KEEP THE NEGATIVE YOU SEE HERE; ADD ANOTHER NEGATIVE FROM THE -10!

$-(\mathbf{-10}) + (\mathbf{5}) = \boxed{15}$

$a = -10 \quad b = 5$

KEEP ON THE LOOKOUT FOR DOUBLE NEGATIVES!

YOU HAVE ONE NEGATIVE SIGN ALREADY, THE -5 ADDS ONE MORE... ...SO YOU REALLY HAVE TWO NEGATIVES IN FRONT OF THE 5!

**2.**   $a - b$

$(\mathbf{-2}) - (\mathbf{-5}) = \boxed{3}$

$a = -2 \quad b = -5$   Double negative turns to addition

**4.**   $-x + (-y)$

$(+4)$

$-(\mathbf{9}) - (\mathbf{-4}) = \boxed{-5}$

$x = 9 \quad y = -4$

**A LITTLE MORE CHALLENGING . . .**

WHEN THEY TOUCH, THEY TIMES!

**5.**   $x + y - z$   $(+12)$

$(\mathbf{-7}) + (\mathbf{-3}) - (\mathbf{-12}) = \boxed{2}$

$x = -7 \quad y = -3 \quad z = -12$

**6.**   $(a)(b + c)$

$(\mathbf{-4})(\mathbf{-2} + \mathbf{-5}) = \boxed{28}$

$a = -4 \quad b = -2 \quad c = -5$

**INTEGER MULTIPLICATION REVIEW**

**7.**   $(-1)(-1) = \boxed{1}$

**10.**   $8 \cdot 6 \cdot (-1) = \boxed{-48}$

**8.**   $(-1)(-1)(-1) = \boxed{-1}$

**11.**   $(-1)(7)(-9)(-1) = \boxed{-63}$

**9.**   $(-1)(-1)(-1)(-1) = \boxed{1}$

**12.**   $(-1)^5 = \boxed{-1}$

5 NEGATIVE ONES ARE MULTIPLYING EACH OTHER!

© Peter Wise, 2014

43

**Add or subtract the following integers**

SSR = Same Sign Rule    OSR = Opposite Sign Rule

SAME sign numbers - you always ADD → SSR
OPPOSITE sign numbers - you always SUBTRACT → OSR

Just give your answer the same sign as the number you added!
Here give your answer the sign of the higher number!

© Peter Wise, 2014

1. $-1 + (-2) =$ **-3**   (SSR)  OSR
2. $-3 - 4 =$ **-7**   (SSR)  OSR
3. $-2 + 7 =$ **5**   SSR  (OSR)
4. $7 - 2 =$ **5**   SSR  (OSR)
5. $5 - 6 =$ **-1**   SSR  (OSR)
6. $-8 - 9 =$ **-17**   (SSR)  OSR
7. $-2 - 4 - 2 =$ **-8**   (SSR)  OSR
8. $-10 + (-10) + (-10) =$ **-30**   (SSR)  OSR
9. $-3 - 3 + 3 + 3 =$ **0**   SSR  OSR  (BOTH)
10. $- -6 =$ **6**
11. $- - -6 =$ **-6**
12. $- - - -6 =$ **6**

HINT! EVERY TWO NEGATIVES CANCEL EACH OTHER OUT!

SHORTCUT! AN *EVEN* NUMBER OF NEGATIVES = *POSITIVE* NUMBER, *ODD* NUMBER OF NEGATIVES = *NEGATIVE* NUMBER

44

---

**Add or subtract the following integers**

1. $5 - 6 =$ **-1**
2. $-5 - 6 =$ **-11**
3. $-10 - 4 =$ **-14**
4. $-10 + 4 =$ **-6**
5. $-10 - 10 =$ **-20**
6. $-10 + 10 =$ **0**
7. $16 + (-17) =$ **-1**
8. $4 + (-8) =$ **-4**
9. $-30 + 20 =$ **-10**
10. $-16 + (-16) =$ **-32**
11. $-2 - 8 =$ **-10**
12. $7 + (-9) =$ **-2**
13. $6 + (-3) =$ **3**
14. $-10 + 12 =$ **2**
15. $9 + (-10) =$ **-1**
16. $-9 + (-10) =$ **-19**
17. $-100 + 4 =$ **-96**
18. $-100 - 4 =$ **-104**

© Peter Wise, 2014

45

---

**Perform the following computations—watch the signs!**

1. $-3 + (-3) =$ **-6**
2. $3 \cdot (-3) =$ **-9**
3. $8 - 12 =$ **-4**
4. $-4 \cdot (3) =$ **-12**
5. $-16 - 4 =$ **-20**
6. $-35 \div 5 =$ **-7**
7. $-7 \cdot 5 =$ **-35**
8. $12 \div (-3) =$ **-4**
9. $-6 + (-3) =$ **-9**
10. $-28 \div (-4) =$ **7**
11. $9 \cdot (-3) =$ **-27**
12. $12 + (-2) =$ **10**
13. $4 - 10 =$ **-6**
14. $-7 + (17) =$ **10**
15. $6 \cdot (-5) =$ **-30**
16. $-80 \div 8 =$ **-10**
17. $-3 + 7 =$ **4**
18. $100 \div (-50) =$ **-2**

© Peter Wise, 2014

46

---

**Perform the following computations—watch the signs!**

1. $-7 \cdot 3 =$ **-21**
2. $-7 \cdot (-3) =$ **21**
3. $(-9) + (-9) =$ **-18**
4. $(8)(-4) =$ **-32**   WHEN THEY TOUCH THEY TIMES!
5. $-16 \div 8 =$ **-2**
6. $14 - 5 =$ **-9**
7. $-6 \cdot 7 =$ **-42**
8. $18 - 20 =$ **-2**
9. $-20 \div (-4) =$ **5**
10. $5 - (-6) =$ **11**
11. $-5 - (-6) =$ **1**
12. $40 \div (-4) =$ **-10**
13. $15 + (-3) =$ **12**
14. $15 \div (-3) =$ **-5**
15. $(-7)(-7) =$ **49**
16. $64 \div (-8) =$ **-8**
17. $-6 \cdot (-3) =$ **18**
18. $(-2)(-3)(-4) =$ **-24**

© Peter Wise, 2014

47

## Fact Families with Negative Numbers

**Example**

| | | MULTIPLICATION | DIVISION |
|---|---|---|---|
| A. | (-3, 4, -12) | $-3 \cdot 4 = -12$ | $-12 \div -3 = 4$ |
| | | $4 \cdot (-3) = -12$ | $-12 \div 4 = -3$ |

switched   same    same   switched

**Give the four fact families for each set of integers**

**1.** (3, -8, -24)

multiplication
- $3 \cdot (-8) = -24$
- $-8 \cdot 3 = -24$

division
- $-24 \div 3 = -8$
- $-24 \div (-8) = 3$

**2.** (-6, 12, -72)

multiplication
- $-6 \cdot 12 = -72$
- $12 \cdot (-6) = -72$

division
- $-72 \div (-6) = 12$
- $-72 \div 12 = -6$

**3.** (-7, -8, 56)

multiplication
- $-7 \cdot (-8) = 56$
- $-8 \cdot (-7) = 56$

division
- $56 \div (-7) = -8$
- $56 \div (-8) = -7$

**4.** (6, -9, -54)

multiplication
- $6 \cdot (-9) = -54$
- $-9 \cdot 6 = -54$

division
- $-54 \div 6 = -9$
- $-54 \div (-9) = 6$

© Peter Wise, 2014

48

---

## Concept Quiz

1. If you have an EVEN number of negative signs in a multiplication or division problem the answer will be (positive) negative. (circle the correct answer)

2. Why does this happen? **Every two negative signs cancel each other out**

3. If you have an ODD number of negative signs in a multiplication or division problem the answer will be positive (negative) (circle the correct answer)

4. One way to think of division problems is to view them as **backwards** multiplication.

5. Fill in the following chart for the fact family for (-28, 7, -4)

| multiplication | division |
|---|---|
| $7 \cdot (-4) = -28$ | $-28 \div (-4) = 7$ |
| $-4 \cdot 7 = -28$ | $-28 \div 7 = -4$ |

6. You can think of multiplication of integers as **repeated** addition.

7. Write this as an addition problem: $2 \cdot (-8)$   **$-8 + (-8) = 16$**

8. BONUS: Calculate $(-3)^3$   **$-27$**

© Peter Wise, 2014

49

---

## Basic Equalities

**Example**

A. $x = -7$   Put a solid circle on -7  ●  ← a number that equals x (or another letter)
Label it x

$y = 2$   Put a solid circle on 2
Label it y

(number line from -10 to 10, x at -7, y at 2)

DOT means EQUALS

**Put solid circles (dots) on the number line and label them**

**1.** $x = 5$   $y = -2$

(number line, y at -2, x at 5)

How far is x from zero? **5**   Left? (Right?)    How far is y from zero? **2**   (Left?) Right?
(circle one)

NOTE: DISTANCE IS ALWAYS POSITIVE!

What is the distance between the two points? **7**

**2.** $x = -8$   $y = -4$

(number line, x at -8, y at -4)

How far is x from zero? **8**   (Left?) Right?    How far is y from zero? **4**   (Left?) Right?
(circle one)

What is the distance between the two points? **4**

**3.** $a = -10$   $b = -8$

(number line, a at -10, b at -8)

How far is a from zero? **10**   (Left?) Right?    How far is b from zero? **8**   (Left?) Right?

What is the distance between the two points? **2**

**4.** $r = -6$   $s = 1$

(number line, r at -6, s at 1)

How far is r from zero? **6**   (Left?) Right?    How far is s from zero? **1**   Left? (Right?)

What is the distance between the two points? **7**

© Peter Wise, 2014

50

---

## Basic Inequalities

**Example**

← less than a number   greater than a number →   ○ ← a number that is less than or greater than x (or another number)
does not include this number
use for LESS than or GREATER than

A. $x < 4$   (number line from -10 to 10, hollow circle at 4, arrow left)
x is LESS than 4

1. Put a hollow circle on 4    2. Draw an arrow going left    LESS = LEFT

**Add the integers by drawing arrows on the number line; put hollow circles where they go**

Less than = arrow goes LEFT     Greater than = arrow goes RIGHT

**1.** $x > 2$   (number line, hollow circle at 2, arrow right)

1. Put a hollow circle on 2    2. Draw an arrow going right    GREATER = RIGHT

**2.** $x < -1$   (number line, hollow circle at -1)

**3.** $x > -6$   (number line, hollow circle at -6)

**4.** $x < 5$   (number line)
start at 0

**5.** $x > -4$   (number line)

51

# Greater Than OR Equal To

**Example**

less than a number | greater than a number

● ← a number that equals x (or another letter)

**A.** $x \le -3$

BOTH ARE COMBINED IN THIS SIGN!

$x = -3$ and $x < -3$

"LESS THAN OR EQUAL TO"

less than 3 | equals (includes) 3

$x \le -3$

**Memory Trick**

a. Put a hollow circle ("donut") if the point on the number line "Do-not" equal a number

b. Put a solid circle ("dot") if it "Do" equal a number

## Write inequalities shown by the number line

1. What inequality does this line show?   $x \le$ **2**

2. What inequality does this line show?   **$x \ge -6$**

3. What inequality does this line show?   **$x \le -1$**

4. What inequality does this line show?   **$x \ge 3$**

© Peter Wise, 2014

---

# Greater Than OR Equal To

## Which inequalities do these number lines show?

1. 
Circle the correct answer:   (a) $x < 2$   (b) $x = 2$   (c) $x \le 2$

2. 
Circle the correct answer:   (a) $x \ge -6$   (b) $x = -6$   (c) $x < -6$

3. 
Circle the correct answer:   (a) $x \ge -2$   (b) $x > -2$   (c) $x < -2$

4. 
Circle the correct answer:   (a) $x \le 8$   (b) $x \ge 8$   (c) $x < 8$

## Draw the inequalities

5. Show $x \ge -5$ on the number line:

6. $x > 5$

© Peter Wise, 2014

---

# Mixed Inequalities

## Write inequalities shown by each number line

1. What inequality does this line show?   **$x < 4$**

2. What inequality does this line show?   **$x \le 4$**

3. What inequality does this line show?   **$x > -6$**

4. What inequality does this line show?   **$x \ge 3$**

5. What inequality does this line show?   **$x > 5$**

6. What inequality does this line show?   **$x \ge -1$**

© Peter Wise, 2014

---

# Mixed Inequalities

## Write inequalities shown by the number line

1. What inequality does this line show?   **$x > 2$**

2. What inequality does this line show?   **$x \ge -3$**

3. What inequality does this line show?   **$x \le 4$**

4. What inequality does this line show?   **$x > -8$**

5. What inequality does this line show?   **$x \le -7$**

6. What inequality does this line show?   **$x > 6$**

© Peter Wise, 2014

# Mixed Inequalities

**Show the following inequalities on the number line**

1. Show $x < 6$ on the number line:

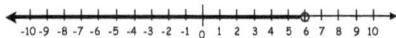

   -10 -9 -8 -7 -6 -5 -4 -3 -2 -1 0 1 2 3 4 5 6 7 8 9 10

2. Show $x \geq 7$ on the number line:

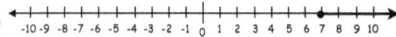

   REMEMBER THAT ON A NUMBER LINE THE = BECOMES A SOLID DOT AND THE > OR < BECOMES AN ARROW!

   -10 -9 -8 -7 -6 -5 -4 -3 -2 -1 0 1 2 3 4 5 6 7 8 9 10

3. Show $x \leq -5$ on the number line:

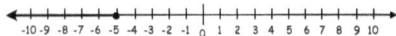

   -10 -9 -8 -7 -6 -5 -4 -3 -2 -1 0 1 2 3 4 5 6 7 8 9 10

4. Show $x > 1$ on the number line:

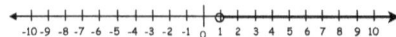

   -10 -9 -8 -7 -6 -5 -4 -3 -2 -1 0 1 2 3 4 5 6 7 8 9 10

5. Show $x < -3$ on the number line:

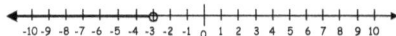

   -10 -9 -8 -7 -6 -5 -4 -3 -2 -1 0 1 2 3 4 5 6 7 8 9 10

6. Show $x \geq -6$ on the number line:

   -10 -9 -8 -7 -6 -5 -4 -3 -2 -1 0 1 2 3 4 5 6 7 8 9 10

56

# Multiplication with Fraction Negatives

**Multiply the following fractions**

**Examples**

**How to multiply fractions: multiply the top, multiply the bottom!**

A. $\frac{1}{2} \cdot \frac{1}{-6} = \boxed{\frac{1 \cdot 1}{2 \cdot (-6)}}$ or $-\boxed{\frac{1}{12}}$

B. $\frac{2}{-3} \cdot \frac{1}{-5} = \boxed{\frac{2}{15}}$

These two negatives cancel out

1. $\frac{2}{3} \cdot \frac{1}{-7} = \boxed{-} \boxed{\frac{2}{21}}$

   USE THIS BOX WHENEVER THE FRACTION IS NEGATIVE!

**Count the number of negative signs in both fractions (numerators and denominators)**

A. If you have an EVEN number of neg. signs, the fraction is positive, because every two negatives cancel each other out!

B. If you have an ODD number of neg. signs, the fraction is negative, because after canceling every two negative signs, you still have one left!

2. $\frac{-5}{6} \cdot \frac{1}{3} = \boxed{-} \boxed{\frac{5}{18}}$

3. $\frac{-2}{9} \cdot \frac{-1}{5} = \boxed{\phantom{-}} \boxed{\frac{2}{45}}$

4. $\frac{-3}{5} \cdot \frac{2}{7} = \boxed{-} \boxed{\frac{6}{35}}$

5. $\frac{2}{7} \cdot \frac{4}{-9} = \boxed{-} \boxed{\frac{8}{63}}$

6. $\frac{7}{-8} \cdot \frac{1}{-3} = \boxed{\phantom{-}} \boxed{\frac{7}{24}}$

7. $\frac{-2}{3} \cdot \frac{-4}{-5} = \boxed{\phantom{-}} \boxed{\frac{8}{15}}$

8. $\frac{-1}{-2} \cdot \frac{-3}{4} = \boxed{-} \boxed{\frac{3}{8}}$

9. $\frac{-2}{11} \cdot \frac{3}{-5} = \boxed{\phantom{-}} \boxed{\frac{6}{55}}$

10. $\frac{-3}{-1} \cdot \frac{-5}{7} = \boxed{-} \boxed{\frac{15}{7}}$

57

# Negative Numbers with Exponents

**Examples**

**IF THE BASE NUMBER IS NEGATIVE. . .**

**Even exponent = positive answer**

IF THE EXPONENT IS AN EVEN NUMBER THE RESULT IS ALWAYS POSITIVE!

$(-5)^2$

$\boxed{-5} \times \boxed{-5} = \boxed{25}$

THESE TWO NEGATIVES TURN TO A POSITIVE!

**ODD exponent = negative answer**

IF THE EXPONENT IS AN ODD NUMBER, THEN THE PRODUCT WILL BE NEGATIVE!

...BECAUSE AFTER YOU CANCEL EVERY TWO NEGATIVE SIGNS, ONE WILL BE LEFT OVER!

$(-2)^3$

$\boxed{-2} \times \boxed{-2} \times \boxed{-2} = \boxed{-8}$

With multiplication and division every two negatives signs cancel each other out!

One negative factor is left over...

...so the answer is negative!

**Strange but true...**

-5 is the same as (-1)(5) ...so -5 × -5 can also be written:

A NEGATIVE SIGN IS THE SAME AS (-1) MULTIPLYING THE POSITIVE NUMBER!

$-a = (-1)a$

With multiplication order doesn't matter so you can rearrange the factors!

$\boxed{(-1)(5)} \times \boxed{(-1)(5)}$

$\boxed{(-1)(-1)} \times \boxed{(5)(5)}$

$(+1) \times (25)$

Every factor is positive, so the answer is positive!

**Multiply the following negative numbers with exponents**

Hint: Just multiply as though the base numbers were positive and then decide the sign!

1. $(-1)^2 = \boxed{1}$

2. $(-1)^3 = \boxed{-1}$

3. $(-2)^2 = \boxed{4}$

4. $(-10)^2 = \boxed{100}$

5. $(-10)^3 = \boxed{-1000}$

6. $(-2)^3 = \boxed{-8}$

7. $(-6)^1 = \boxed{-6}$

8. $(-7)^2 = \boxed{49}$

9. $(-1)^{15} = \boxed{-1}$

58

# Intro to Absolute Value

**Examples**

Absolute value is always POSITIVE (or zero); NEVER NEGATIVE

Absolute value is the DISTANCE from ZERO (which of course can never be negative)

The absolute value of a number is represented by two parallel, vertical lines $|3|$

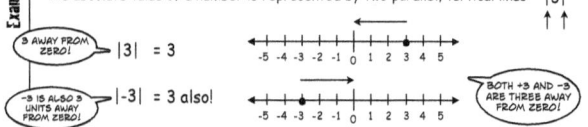

3 AWAY FROM ZERO!

$|3| = 3$

-5 -4 -3 -2 -1 0 1 2 3 4 5

-3 IS ALSO 3 UNITS AWAY FROM ZERO!

$|-3| = 3$ also!

-5 -4 -3 -2 -1 0 1 2 3 4 5

BOTH +3 AND -3 ARE THREE AWAY FROM ZERO!

**Answer the following absolute value problems**

1. $|-7| = \boxed{7}$

   Just make the value inside positive

2. $|8| = \boxed{8}$

   Just keep the value positive

3. $|-2| = \boxed{2}$

4. $|-a| = \boxed{a}$

   (assuming that $a \geq 0$)

5. $|9| = \boxed{9}$

6. $|-12| = \boxed{12}$

**A LITTLE HARDER...**

7. $|1-5| = \boxed{-4} = \boxed{4}$

   SOLVE THE PROBLEM INSIDE FIRST, THEN MAKE YOUR ANSWER POSITIVE!

8. $|-2 -7| = \boxed{-9} = \boxed{9}$

9. $|10 + (-13)| = \boxed{-3} = \boxed{3}$

10. $|(-3)(4)| = \boxed{-12} = \boxed{12}$

11. $|(-2)(-2)(-5)| = \boxed{-20} = \boxed{20}$

59

## Absolute Value and the Opposite Sign Rule

Aha! Now that you know about absolute value you will learn some interesting things about integers! Think about the sign rules!

| Normal Subtraction | Absolute Value | |
|---|---|---|
| 5 - 2 = | \|5 - 2\| = | |
| 1a. **3** | 1b. **3** = **3** | |
| 2 - 5 = | \|2 - 5\| = | WHAT DO YOU NOTICE? |
| 2a. **-3** | 2b. **-3** = **3** | |

| 10 - 1 = | \|10 - 1\| = | 4 + (-3) = | \|4 + (-3)\| = |
|---|---|---|---|
| 3a. **9** | 3b. **9** = **9** | 5a. **1** | 5b. **1** = **1** |
| 1 - 10 = | \|1 - 10\| = | -3 + 4 = | \|(-3) + 4)\| = |
| 4a. **-9** | 4b. **-9** = **9** | 6a. **1** | 6b. **1** = **1** |

OBSERVATIONS: What is the relationship between a - b and b - a?

**The answers have the same absolute value**

### Practice with the following absolute value problems

| \|2-6\| = | \|8 + (-10)\| = | \|7 + (-4)\| = |
|---|---|---|
| 7. **-4** = **4** | 8. **-2** = **2** | 9. **3** = **3** |

60

---

## Practice with Absolute Value

**Absolute value**
- Measures distance from zero
- Always positive (or zero); never negative

### Use your knowledge of ABSOLUTE VALUE to solve the following problems

1. \|-5\| = **5**
2. \|-7\| is **7** units away from zero
3. \|-12\| + \|-4\| = **16**
4. \|-10\| - \|-2\| = **8**
5. \|-a\| = **a**  (assuming that a ≥ 0)

6. \|-x\| + \|-x\| = **2x**  (assuming that a ≥ 0)
7. \|-4\| × \|-6\| = **24**
8. \|-20\| × \|-2\| = **40**
9. $\left|-\frac{2}{3}\right|$ = $\frac{2}{3}$
10. \|-15\| × \|-10\| = **150**

### Order these values from LEAST to GREATEST

11. \|-12\| \|-4\| \|-1\|   *(LEAVE YOUR ANSWERS IN THE SAME ABSOLUTE VALUE AS ABOVE!)*

**\|-1\|  \|-4\|  \|-12\|**

12. \|-6\|  5  \|-8\|

**5  \|-6\|  \|-8\|**

13. \|-10\| \|-2\| 18

**\|-2\|  \|-10\|  18**

14. $\left|-\frac{1}{4}\right|$  $\left|-\frac{1}{5}\right|$

$\left|-\frac{1}{5}\right|$  $\left|-\frac{1}{4}\right|$

61

---

## Absolute Value and the Same Sign Rule

Aha! Now that you know about absolute value you will learn some interesting things about integers! Think about the Sign Rules!

| Normal Addition/Subtraction | Absolute Value | |
|---|---|---|
| -2 - 2 = | \|-2 - 2\| = | |
| 1a. **-4** | 1b. **-4** = **4** | |
| 4 + 4 = | \|4 + 4\| = | WHAT DO YOU NOTICE? |
| 2a. **8** | 2b. **8** = **8** | |

| 6  6 = | \|-6 - 6\| = | -10 + (-5) = | \|-10 + (-5)\| = |
|---|---|---|---|
| 3a. **-12** | 3b. **-12** = **12** | 5a. **-15** | 5b. **-15** = **15** |
| -3 + (-5) = | \|-3 + (-5)\| = | -4 - 6 = | \|-4 - 6\| = |
| 4a. **-8** | 4b. **-8** = **8** | 6a. **-10** | 6b. **-10** = **10** |

OBSERVATIONS: What is the relationship between a + b and -a + (-b)?

**The answers to both have the same absolute value**

### Practice with the following absolute value problems

| \|6 + (-8)\| = | \|-7 - 7\| = | \|15 + (-3)\| = |
|---|---|---|
| 7. **-2** = **2** | 8. **-14** = **14** | 9. **12** = **12** |

62

---

## Integer Test

1. 4 - 10 = **-6**
2. -8 - (-4) = **-4**
3. -12 - 7 = **-19**
4. 5 + (-7) + 8 + (-3) = **3**
5. -13 + 9 + (-8) + 5 + (-6) = **-13**
6. 7 · (-9) = **-63**
7. (-2)(-2)(-2) = **-8**
8. -24 ÷ (-4) = **6**
9. 35 ÷ (-7) = **-5**
10. 6x + (-14x) = **-8x**
11. -9x + (-7x) = **-16x**
12. - (-7) - 6 = **1**
13. 7 + **-14** = -7

14. **-12** - 8 = -20
15. -x + (-y) = **-7**    x = 12   y = -5
16. $(-9)^2$ = **81**
17. \| -8\| = **8**
18. \|(-4)(7)\| = **28**
19. Order these from LEAST to GREATEST

    \|-14\| \|-6\| \|-2\|

    **\|-2\|  \|-6\|  \|-14\|**

20. Show x < -3 on the number line:

21. Show x ≥ -5 on the number line:

63

Now that you've learned the basics of integers, it's time to move on to the next level with

# Algebra, Book One

**Topics include:**

- Basic Algebra Terms
- Introducing Variables and Coefficients
- Adding and Subtracting Variables
- Adding Variables with Exponents
- Substitution
- Turning Words into Algebra Symbols
- Like Terms
- Undoing Addition and Subtraction
- Undoing Multiplication and Division
- Solving 2-Step Equations
- Exponents = Number of Repeated Factors
- Variables with Exponents
- Multiplying Numbers and Letters
- Adding vs. Multiplying Variables
- Dividing Same Bases with Exponents
- Getting x Only on One Side
- Fraction Coefficients
- The Distributive Property with Variables

## ...with more tips, tricks, and wackiness!